DEAR HUMANS

NAZARENO FABBRETTI

Animals write...
DEAR HUMANS

Translated by
Deborah Misuri Charkham

Original title: *Caro uomo (Lettere degli animali)*

Illustrations: pages 30, 36, 55, 73, 80, 85, 121, 129 by Gino Gavioli; pages 43, 49, 68, 91, 108, 115, 137 by Mary Lou Winters

St Paul Publications
Middlegreen, Slough SL3 6BT, United Kingdom

ISBN 085439 349 8
Printed by Billing & Sons, Worcester

St Paul Publications is an activity of the priests and brothers of the Society of St Paul who proclaim the Gospel through the media of social communication

CONTENTS

I find it hard to believe
that the rustling sound of wind in the leaves
is not an oracle;
hard to believe that this dog,
my brother, has no soul.
Simone Weil

Animals are the smallest part
of divine creation,
but we will see them again one day
in Christ's mystery.
Paul VI

INTRODUCTION

Animals teach us a new lesson
by Francesco Alberoni

Nowadays human beings are doing all they can to go back to nature with a greater understanding and to rebuild the bridges which seemed burnt. For many centuries, and particularly with industrial development, humans have gone against nature by trying to dominate, enslave and change it. Nature was considered to be a hostile force to be conquered by imposing human order. The entire earth had to be explored, criss- crossed with roads and railways. The virgin forests had to be transformed into agricultural land. By competing in this way, animals were placed in two distinct categories: those which were domesticated and useful to the human race, and the rest which were wild, hostile and dangerous.

Attitudes changed with the realisation that our life on the planet also depends on the delicate ecological balance. Conditions for survival of the human race are the same as those governing other living beings, our fellow-citizens on earth. The abyss separating humans from animals was made narrower during the last century by the theory of evolution. However, more recent biological studies have underlined the importance of the behaviour of living things. In his book, *The Life of Life*, Edgar Morin supports the theory of subjectivity. Being a subject is a property of all living things, beginning with the cell. Every living being "earns a living" in a risky process of self-building where mistakes can always be made and, with the mistakes, comes death. Every living being must "look after himself", battling for his own survival. At the same time, he "takes care of others", that is to say the multicellular organism of which he is part, the herd or, like the ant, the bee and the human, the community.

Another condition common to all living beings is the need to grow and feed off other living beings. Plants live on humus produced by the decomposition of other plants, whereas animals usually feed off living beings and many of them on other animals which suffer, try to escape and are afraid.

The main characteristic of superior animals, especially mammals, is emotional feeling. Morin writes: "The category of mammals is the culture medium for affection. The mammal is a warm machine. Inter-relationships between mammals are deeply emotional... We mammals are beings interwoven with pathos. Pathos is our very existence." Pain, tension, joy, pity, love, attachment, anxiety, euphoria, fun: all these things are common to us and our animal brothers whose company we enjoy but, more often than not, whom we kill for their skins or meat.

Humans share the natural condition of the animal in every way. Humans grow by using all the other living beings to "earn a living". We live, thanks to their death. We live by taking their lives, in exactly the same way they do. There is no other choice.

Perhaps this human condition, condemned to kill and cause suffering, is that which theologians once called original sin. It is a condition in which the human cannot help but cause pain. We begin at the very moment we want to live, and we never stop until the moment we die. I wonder whether the gnostics and Cathars were so pessimistic about the world because they recognised this aspect of existence and understood that there was no way to escape. It seemed impossible to them that a benevolent god could have created a world where beings live in continual anguish about life and are forced to kill each other in order to survive. And so they thought that the world, this world, must have been created by a malevolent god. Through their moral conduct, they did everything they could to put a stop to this life, in this way making room for the god of light and his kingdom.

With their beauty, grace and sensitivity, and at the same time with their relentless cruelty, animals present us with the worrying problem of pain and death. They make us see ourselves as we really are. We see our insecurity and impotence reflected in them. We discover the failure of our moral intentions and our limited capacity for redemption in these, our inferior brothers. Whatever progress we might make, we will never be able to avoid death. Even if one day we manage to prolong human life, we will never be able to do the same for all the other living creatures we need for our survival.

When Nazareno Fabbretti says that, as a child, he believed that animals also went to heaven, he is expressing this desire for human redemption to be extended to the natural world at large, aspiring to a form of universal being unimpeded by death and evil. Since Christ, for the Christian, came to save the human race, the destiny of animals and other living beings remains a mystery. Why does salvation and redemption only apply to the human race? Why is only human suffering important, and not the silent suffering of other creatures? St Francis talked to the birds and, with loving words, convinced the wolf of Gubbio to mend his ways. Are they, therefore, not equally important?

When I think of the founders of modern morals, Jeremy Bentham, the creator of utilitarianism, springs to mind. According to Bentham, the needs, feelings, aims and aspirations of living beings are the only things that matter. For both humans and animals. And the lesson to be learned is how to choose the actions which will create the most happiness for the majority. According to Bentham, since animals also suffer we have a moral duty towards them. We must reduce their suffering in every way we can and increase their happiness. Bentham's moral teachings have no need for religious back-up and have nothing to do with superior human rank. However, because of their very simplicity, they can also be immediately applied to our inferior brothers.

Unlike us humans, animals do not have an awareness of death. They know fear, anguish and suffering but have no awareness of death. Human suffering is essentially based on an awareness of mortality. No-one can really comfort the human, no-one can give him eternal hope. Because there is death. This obstacle does not exist for the animal. We can feel sorry for them, for their death. But they don't feel this because they don't know and will never know. As a result, we can do a lot for them. As long as we take the moral lesson Bentham has taught us literally: make sure they are happy, treat them properly, understand and satisfy their needs. Don't make them suffer. If we must kill them, make sure they don't suffer fear or pain.

This moral conduct is perhaps not enough for us. But it is enough for them. From their point of view, there's no need for anything else.

FOREWORD

Mea Culpa

I confess my sin of arrogance in these pages. But I hope I will be forgiven because they are written with sincerity.

Without being a naturalist or a writer of fables of any merit, I have taken it upon myself to call upon some animals to take the stage and address you directly. I imagined each animal writing a letter to persuade humans to help them avoid their collective disappearance from the planet. I reached the point of using them as my mouthpiece to present my own thoughts and feelings through their words. Because the main aim of this book was one of provocation, in each individual case I realise that I have behaved worse than the animals themselves, and have in many ways distorted them by giving them a little too much human aggression.

However, there is also a personal motive which justifies these pages and the metaphor and symbols to which the animals are subjected. All the large and small domestic pets with whom I have shared my life since I was small, and all the others I have come across in the world, both free and in captivity, have always caused me to feel both love and anguish. In one way or another, I viewed them all as predestined victims. Very rarely did I come across truly free animals even if they weren't in cages, laboratories or abattoirs. I realised that the most difficult problem for them, as it is for us, is freedom.

As a child, I wanted to free every animal I came across whether they were in a cage or roaming around my home. The only animals I considered to have freedom of choice were a dog and a few cats who insisted on returning to our house even though they had been sent away. But were they really free to choose?

Although I often wasn't aware of it, I learned a great deal from them: the right instincts, freedom of thought, a sense of friendship and the suffering of being forced to feign unspontaneous obedience. Since childhood I have always believed that "animals also go to heaven" because, in so many ways, they already have hell and purgatory right here.

I was always teased by my family and friends because of my faith in "resurrection" and "immortality" for animals. Later on, when I actually wrote about it, the only reaction was either gentle condescension or cutting sarcasm. But I never changed my mind. Many years later, when I was no longer young but more than ever convinced of this belief, I was finally compensated when I least expected it by the simple and enthusiastic words of the great and very sensitive Pope Paul VI. One day he said that animals "are the smallest part of divine creation, but we will see them again one day in Christ's mystery". Another pope, John Paul I, a good writer of fables, defined them with a term used by Christ when referring to the poor "our smaller brothers", and therefore worthy of fraternal affection and solidarity.

I had this in mind at the time when I wanted to show some sign of this solidarity by subscribing to ecological, environmental, anti-vivisectionist and pacifist movements. In addition to the biblical condemnation for humans to eat their flesh, my tolerance for this outrageous death sentence on animals was further reduced by their suffering at the hands of hypocritical scientific games like vivisection, in the methods of slaughter, by being "thrashed out" to be hunted, through artificial insemination, intensive fattening and death.

The book was created in the certainty that there can be no illusions about planetary reconciliation between humans and animals, and also in the conviction that we must fight against this inhuman "intellectual pessimism" and put the "optimism of good will" into action. Many are beginning to understand, but only very slowly.

I fear that the sun will set on our post-industrial and technocratic generations before the choices in defence of the animals have become a practical and practicable way of life. No matter how difficult to keep alive it may seem, extinguishing this flicker of hope would mean increasing and accelerating both the aggression and fragility of the human and animal species, pushing towards the silent "big bang", the breaking of anthropological "links" and the environmental balance.

I have tried to hear and understand what the animals think about this risk. Even if Elsa Morante has written that "no word in the human language could comfort a guinea-pig", I have carried on raising questions about both my childhood love of all animals and the evolution of their language addressed to us. This language must surely exist and contain many messages. I know that by humanising them far more than possible, I have been far too presumptuous with my "anthropomorphism" of the hypothetical authors of these letters. But, unless I used my imagination – the highest form of real faith – in a subject like this there was no other way to come to terms with these messages, these ultimatums. It is a show of respect, as well as love, for creatures which, like us, are educationalists and teachers of life for us all.

Thus, by writing I found myself listening.

"Dear Humans..."

Nazareno Fabbretti

THE LION

Dear Humans,

I am fed up with this crown. I will probably disappear from the face of the earth before ever understanding why you have given it to me. I never asked for it and I don't quite know what to do with it. King of the Animals, King of the Jungle: can you tell me what this really means?

I would like to abdicate. Not, however, to be replaced by another animal, but to be replaced by you, a human. After all it was you who landed me with this title, without even asking yourself what it might mean for me or for the other animals. As you appear to be trying to exterminate the animal kingdom, why don't you take up this farcical title? After all, you are always aspiring to be king and boss of someone or something, to dominate everyone and everything.

What exactly did God tell you men and women when he gave you life by creating your ancestors Adam and Eve, and joyfully prepared that great garden full of delights for you? Did he say "dominate the earth" or did he say "take care of the garden"? Both versions are contained in the first book of your history and faith but, since the beginning of your existence, you have always preferred "dominate the earth, eat all living things contained there". Am I wrong?

What underhanded reason did you have for so hastily assigning me the title of King of the Animals, King of the Jungle? You, dear human, you are the king. You have turned the whole marvellous world and universe into a "jungle" and only you can take responsibility and be king. King of all "animals", of all beings with the faintest glimmer of a soul.

I am only a fine, strong beast; nothing more. There is nothing else to distinguish me from the other animals. Animals don't have kings. All we have are leaders of our herds and their role is to guide and defend their respective groups. There are no dynasties and no crowns. We all have our own stature, strength and intelligence. It is true to say that we also have our ferocity, but we need this to survive. The elephant and other giant animals mightier than myself are without doubt the strongest and even the most intelligent. The little monkey is the most amusing; the tiger the most agile and fierce; the horse is as handsome as I am, if not more so.

You, humans, are neither my slaves nor my masters. During the long rest periods in the shade, between hunts for the fresh meat I need for my survival, I always ask myself what you mean to me and why, for centuries, you have stubbornly insisted on placing this symbolic crown on my head. Is it because you need me to be great, handsome, strong and more regal than I really am so that you can have an adversary, a target, prey worthy of your ambitions and megalomania?

You weren't just content to hunt me down all over the world until I almost became extinct. You have also given me a starring role in your celebrative literature which seems to please human adults so much and, as a result, is also enjoyed by your children. Your stories and novels about "big game hunting" – such as those written by the proud and chivalrous Kipling, or the shrewd and insensitive Salgari – make me the protagonist of a farce. But the farce always turns into a bloodthirsty tragedy full of hypocrisy and lies, where you are the true protagonist because you are the clever, stylish killer. And, when you have killed, you want your proud prey to be admired.

Even your most insignificant kings have taken an interest in me and crucified me on their coats of arms, their flags and their crowns. I have become a symbol, an analogy, an emblem of strength and nobility. I have helped comfort them for their frailty by making them appear to be as strong and invincible as I am. Even the Christians among you, in your Old and New Testaments, have stolen my image and used it as a symbol to adorn your flags, to fortify your armies and conquer. However, what you most needed to conquer were your own fears. This is how a dark-skinned king became known as the "Lion of Judah". But "Lion of Judah" is also a Christian prayer to ward off the devil, a prayer attributed to the peace-loving Anthony of Padua: "The devil's powers must flee; the lion of the tribe of Judah shall conquer". And that is to say, Jesus Christ. When Saint Peter, the peaceful fisher of fish and people, wanted to give some idea of the devil's force, he compared him with a lion: "Satan, like a roaring lion, will roam among you to devour you."

I am not offended by these comparisons. Many of them were made in good faith. The thing that does bother me, however, is that you always bring up the matter of my strength and cruelty to give yourselves prestige and honour. It is quite true that I am strong, but the bit about my cruelty is slanderous folklore. When you, the human, tuck into a delicious lamb chop, do you consider yourself cruel or simply hungry?

I hunt and kill and shed blood only when I am hungry. You humans say so yourselves in your books about the animal world, but you soon forget what you have said when you arm yourselves to hunt frantically for alibis to blame me for all your secret yet obvious frustrations. After all, who isn't cruel? Apart from you and me, aren't the insect and the tiger also cruel in their own way?

If your zoologists and naturalists had correctly recorded my habits and my code of conduct towards other animals, they would have always taught you from a very young age that I am never the first to attack other animals or humans, and especially not without good reason like hunger or fear. Don't flatter yourselves: apart from anything else, we don't even like the taste of your meat and blood. In short, we consider you better than nothing in a dire emergency – not very satisfying nourishment. For me, there is nothing more delicious than a lamb or a gazelle. There are times when I have even considered becoming a vegetarian.

However, hunting is still the lesser of your crimes. Even if we don't use the same weapons, we are at least even on this score.

You commit your greatest offence with your safaris and circuses. To be quite honest, the circus is even more humiliating than the safari. Better to die than be endlessly piled up, precariously balanced, performing cruel and stupid tricks to entertain you and your children. When we compare the two options, we prefer those horrible rows of embalmed lion heads decorating the walls of your most ill- fated and exhibitionist hunters.

You may say that amusing thousands of pensioners and children in a circus tent is a fine thing to do. However, these are your words, not mine. This is your excuse for making money even if you try to use entertainment as a reason to clear your conscience.

Don't you realise that we totally lose our identity in your circus? Food, space, behaviour, relationships: you want all these things to be "human" and not "animal". Your dramatic whipping of frustrated, and often drugged, animals not only revolts us but also fills us with such terror that we have become morally subjugated and tamed. I must confess, this horrifies me so much that, at times, I instinctively feel very happy to know that a lion, in a passing flash of regained dignity as well as desperation and hunger, has injured the tamer who had been presumptuous enough to risk his life by sticking his head into the lion's mouth. In their own way, those lion-tamers – poor devils, carrying whips and dressed up as clowns – are, like us, victims of collective alienation.

I can speak for all of us when I say that it would be wonderful to be able to entertain your pensioners and children. But only if we become friends once more. I do realise, however, that both the animal and the human will have to find something completely new deep within us in order for us to return to the way we used to be before that great (or small?) mistake which Adam and Eve made. It created a dark and violent destiny for us all. The darkness which fell on all living creatures immediately after the first glorious sunrise needs to be dispersed.

Don't you think that the most beautiful and exciting adventure you and I could have would be to find, rebuild and relive that distant, sincere and happy friendship?

Seven hundred years before Christ, one of your greatest poets, the prophet Isaiah, caught sight of the possibility, the certainty in fact, of this miracle. He said that the day would come – with the birth of our brother Jesus Christ – when everyone, animals and people, would become friends and brothers once more. The wolf and the lamb would sleep side by side, the child would happily play with the snake, the lion and the human would talk as friends. War would end; there would be no more hunting; there would be other foods, other games, other animals, other people. And a little boy would guide them. Yes, this is what your prophet said: "A little boy will guide them". Don't you think this is marvellous, humans, you old battle-axes?

No more massacres, no more "shows". It may be buried deep, but you and I, your people and my species, have this friendship within us. All the creatures in the world carry this sparkling secret.

Don't make lions into men and don't make men into lions. There has already been an emperor in Africa who liked to call himself the "Lion of Judah" and his dynasty has disappeared. The great British monarchy, which built an immense empire, has a lion on its coat of arms but, luckily, the empire now only exists in the form of friendship. Learn from history, dear humans, and think.

Go back to "nature", live more simply, don't overindulge in images and brutal myths of power. Go back, spontaneously and gradually, as friends not as hunters and tyrants, go back to to the countries and places where we still survive. You can save our species by rehumanising your own. In time you will see your children playing with our cubs, imitating our God, your God and mine, who, at the beginning as the Bible says, "played" on earth himself. Even though he was God, perhaps he wanted to become a child like other children so that he could enjoy himself more in the games. This is the reason why I want to give up the crown – this useless, undeserved symbol of power which you have given me. All symbols of power – before, during and after they have been used by humans and beasts – signify struggle, blood, massacre and death.

Humans, dear humans, brother humans, please don't be offended. Here is your crown back. It is not mine, it's yours. It is worthy of you and you of it. For better or for worse.

THE MONKEY

Dear Humans,

hese words were written by one of your scholars who spent his whole life studying monkeys and knows us well: "If monkeys were to get bored, they would become human beings".

Don't worry, humans. We monkeys will never get bored and we'll never become human beings. Grandma Chimpanzee says that this is jolly lucky for us, and maybe it is for you too. You have never liked the idea that we could have been your ancestors and are far too ashamed to accept the hypothesis now.

We are monkeys and you are humans. We both have our place, and our relationship will be more dignified and more sincere if we speak our minds.

I am one of the last monkeys in the "zoo", to use your word, and I am writing to you on behalf of all of us. I have reached a ripe old age and have seen hundreds and thousands of men and women, both young and old, looking through the wire fence around our enclosures. I call them lagers even if you stubbornly insist on calling them zoological gardens. Except for your children, you are a sad and pitiful sight.

You adults are so hypocritical when you look at us. You find it difficult to hide your disgust although you ought to be cheerful and amused. A monkey? A monkey is funny, always funny. You throw her some peanuts or sweets and she'll shell them or take the wrapper off and quickly munch them or greedily gobble them up. While she's doing this she never takes her sarcastic eye off the donor. Even though she never stops leaping from one side of her fake forest to the other, her only pleasure is the sound of children squealing with joy.

The monkey is, in fact, quite moved by those squeals and willingly performs her tricks. Just for a moment she imagines that she really is back in the forest of her ancestors, meeting Tarzan's children or grandchildren. But Tarzan, although a very agreeable character, is only the imaginary protagonist of your exciting stories and none of us has ever really met him.

As soon as the peanuts and sweets have been finished, the imaginary forest becomes a ridiculous, artificial lager again. When the children and adults have gone home, we still feel your gaze upon us and have never known whether to interpret it as affection or derision, embarrassment or happiness, respect or scorn.

Once again we ask ourselves: is it really true that this human is our descendant? Can it be true that we are the missing link in a chain which ends with you? This is claimed by another of your famous scholars – one of the learned men of science who hands out science and culture in pills. However, we don't waste too much time with these questions. Your science is not ours and there is a great difference in our cultures. We live by our instinct and, judging from the results, our instinct seems to be sharper than your tiresome, proud intelligence. You are proud, while we are just curious. This is where the difference lies. Curiosity amuses, while pride devours.

It's true, you have come back to us, but in the most inhuman and ferocious way. You should be ashamed of the experiments being set up by your most brazen scientists. They have already tried to mate our two species to produce a monster which one of you wanted to call the "chimpanman". Luckily it was unsuccessful.

Not only is this a wild dream, it is an absolute outrage. It's always the same old story: we seem destined to be your "guinea pigs", like dogs and cats, confirming your ambitions which lie somewhere between Satan and Faust.

Aren't you content with the human body, beautifully designed and moulded to perfection by the Creator, the flesh he chose to embody his Son? Or are you so insecure and vulnerable that you have to pretend and hide behind the mask of a monkey for safety, to have greater success with your devastation and pollution, your dreams of plunder and glory? Do you call this progress? Or regress? Isn't coming back to us the same as taking a step backwards? With all your pride, how can you forgive such a step?

Because I understand your boredom, I can also understand your need for constant change, constant search, never satisfied with what you have. This is a sign of your greatness, but also a sign of your unhappiness. I can't stand your boredom because I have always felt brotherly love for you. I fear that you will end up by creating an even more devastating catastrophe, for yourself and for us, because of this boredom and your need to overcome it at any price.

You humans own houses, live in great cities and control the world. The Creator gave it to you and you have fun with it, but more often than not, you are like cruel children having fun torturing a cat or a dog or destroying a toy. Aggression, which we use as a defence mechanism, has become for you "the noble art of the hunt", "the noble art of war", a means to overpower your families or the whole planet. You never pull your punches. Whether you live in a comfortable house in a huge city, or in a hovel or hut, the only pastime which fully occupies you is the overpowering of someone else. The cities and houses offer you the possibility to live together, communicate, love one another, help each other and enjoy yourselves but you live to hate each other, to win, to cast each other out and, in the end, to become totally bored. You lie to each other, avoid each other on crowded trains, in supermarkets, stadiums, on the buses. The real proof of your boredom can be seen in your fear of lifts. Unless you are in a lift with friends or relations, you don't know where to look or what to say. You go up and down in silence, your glazed eyes staring into space. You are monstrous products of indifference and fear.

You have lost your innocence and no longer have the courage to look each other in the eye without fear. You live in a world of insane, absolute rulers. In the cities, and above all in your homes, you are living more and more as strangers. If two strangers were to meet in a desert, I am sure they would greet each other and continue their journey together. The same strangers, in a lift, are divided by a wall of silence and embarrassment. Strangers among strangers. Dear human, think about the lift now and again. I believe that it most reflects your fear and your solitude.

Three cheers for the forest! Even though our life is just as difficult as yours, imagination and friendship are constants in the forest. We know that there are some very good individuals and groups of people among you, conscientiously fighting against the destruction of our environment, the fruit trees, the air, water, the earth. This pleases us and we'd like to give you a hand. But will we be in time to put a stop to the known – and unknown – disaster threatening the balance of the universe?

Fuelled and propelled by this contagious boredom, you are even trying your luck on the moon, and want to go beyond whatever the cost. However, none of you are willing to risk your own lives to test resistance to the effects of the stratosphere. "Guinea pigs" are needed and so you have developed the habit of putting dogs and monkeys into your clever and sophisticated contraptions and launching them into space. Dear human, you get so excited and are so quick to react to anything new, unknown or mysterious. So we have become your explorers, your new space "slaves". And as soon as you discover something, you become bored and start to yawn.

We monkeys were so thrilled about the joke – part accident, part luck – played on you by a little monkey launched into space with her hands tied so she wouldn't touch anything. By accident, or perhaps fortuitously, she was fascinated by all those forbidden buttons, levers, and lights on the control panel of the space ship and wanted to play with them. She managed to free her left hand and got into real mischief. She didn't mean any harm, and needed some fun because she was bored cooped up in there. Like a human, she needed that game to alleviate her boredom. She returned happily to earth, glad to have had some fun, even when the astronautical technicians wasted no time in grabbing hold of her to study her levels of contamination or infection contracted at such heights. Perhaps by now she has already been cut open so her organs can be examined more closely?

Where exactly is your science leading you, human? Which ethic do you follow? In the name of which civilisation do you declare peace and then break your promise? Why are you poisoning the world and the universe? Why lose our friendship?

I realise it's not my place to moralise. I am only writing to you to ease my conscience and because you still have the chance, though you don't deserve it, to use your intelligence and the means at your disposal to save yourselves and us at the same time. A bit like

Noah's Ark, when all living things, humans and animals, were able to escape extinction. You need another ark, my friend. But it has to be bigger and safer than the first one to hold everyone together and not be shipwrecked. You have to deserve the rainbow: the sign of God's pact and friendship with humans.

For this reason you need us animals. Noah built the flood- proof ark but only the dove, an animal, was able to tell him that the flood was abating when it brought back the famous olive branch. Since then the dove and the olive branch have signified peace.

Human, dear human, we are still all in the same boat, either out of love or because we have no choice. And our only hope of salvation is to stick together. When Noah's Ark touched land, we monkeys began to dance and jump for joy. We have never stopped.

Why don't you join us?

THE DOG

Dear Humans,

s far as I know I am one of the few dogs, perhaps the only one, to have miraculously escaped from the cruel practice of vivisection. I spent a long time on the operating table, in the hands of a famous surgeon. Although he was a well-known scientist, to me he was more a specialist – revered and overpaid – in the shameless and continual vivisection of his fellow creatures.

I am writing to you for a precise reason. Whatever the cost, I have decided to found a League of Dogs to obtain the abolition of vivisection because, sooner or later, it will be the dogs who practise it on humans.

I lead a reasonable life these days. I guide an unfortunate young man who lost his only good eye when a top ophthalmologist made a mistake and didn't even bother to ask forgiveness of his patient's family. This boy is good and obedient but he had tremendous difficulty in becoming obedient. I knew why he was so angry at first and we were able to reach an understanding straight away. Now we live more or less as brothers. I suppose you could say that we are ill-fated brothers, but brothers none the less. A few years ago we both got a mention in the papers with a couple of ten-line articles for a couple of days. Then nothing. The boy became very depressed. His parents gave me an artificial limb for the most damaged of my legs, the stump, so that I could walk. But they were old and sick and died soon after.

I haven't written to you before because I had it in for you, for the whole human race. Now, however, I am pleased to be able to quietly tell you what I think, right now, about myself as "man's best friend" and, if you will allow me, about you as the "dog's friend".

On your side, there has always been a huge difference between the way you usually treat a dog and a cat. If I am not mistaken, you have to win the cat over while the dog wins you over. Your heart is the first to melt even though you try to dominate him. Then, if all goes well, you begin to love him and as soon as you love him you become his servant, if not his slave. Don't tell me it isn't so because, if you haven't realised this by now, it means that you really are irretrievable.

I often watch you closely, when you're not looking, and ask myself whether that writer – I can't remember his name – was right when he said that when a human and a dog live together for a long time, they start to look alike. I don't know whether this has happened to us, to me. I have to admit that it would be an advantage for me, but I don't know how you would take it.

So let's take a slightly broader look at the situation.

Without doubt you and I, in a funny sort of way, are the most persistently talked about couple in the world, whether individually or together. Although we are envied, we are also the most misunderstood and slandered couple. The cat, for instance, doesn't like me because of this and who knows what cats go around saying about us.

So, am I or am I not "man's best friend"? For me it would be an honour, a great joy and a life-long ideal. But how do you feel about it? Without a shadow of a doubt you are the master and owner. But are you my friend?

When I was undergoing vivisection and desperately trying to cling on to the need to know why all this was happening, one of the team who had already sliced me up felt as lowly as a worm and had to shout out his feelings to the others (I hope the worm will forgive me for using one of your comparisons). He hated himself and the others for what they were doing. He sewed me up as best he could (it was his job), and, as I was still alive, instead of "putting me to sleep" and dumping me in the garbage with all the infernal human and animal remains, he took me home with him. It was like heaven. The surgeon's ageing parents looked after me with tender loving care, as though I was their child. When it was decided that I should become a guide dog it was as though they had read my mind. Although I had paid a high price, I was almost grateful to the barbarity of vivisection for having given me the opportunity to meet people like this and to have been able to know such love in the home of a slaughterer.

Now it is comforting to feel and know that I am indispensable. The both of us together are like one perfect creature. We are not dependent on each other out of necessity, but out of solidarity and affection. We got on well from the very start. Someone told my blind friend that a film should be made about our story... but heaven forbid! I'd run away immediately. I don't want to make a spectacle of myself and be the cause of crocodile tears. When faced with a true story, my story, they'd all rush off, hurriedly drying their watery eyes.

Our story doesn't have the makings of a film plot, but it definitely carries a message. My blind man already knows it, but I have to tell you personally, and all humans, that I have only done my duty. Everything else has remained the same as it has always been. How long will it take – years or even centuries – for a law to be made against the totally irrational practice of vivisection?

There is no logic in a dog's love, but it is sincere and all-embracing. Rarely does this love diminish, change or disappear even if you make us die of hunger or indigestion.

On the rare occasions when a frightened or starving dog becomes ferocious and attacks a child, all your pretty stories about dogs being warm and friendly fly out the window. You call us wolves instead of dogs and, by doing this, you offend both the dog and the wolf. You know full well that we only become ferocious when we are starved or when we defend someone we love – like you, our friend and master. For you we would challenge someone stronger than ourselves and are prepared to kill or be killed. Our allegiance and love is so great that we often dare to tear apart and kill whoever threatens or attacks you.

In the days before vivisection, I regarded you differently. I saw you as a god. This is where I made my mistake, the mistake made by all loving dogs. My mistake was so great that I even went hunting with you as though I was going to a wonderful, exciting party. It was only later that I understood how you are rich, fed up and bored and go hunting because you have to get rid of your frustrations and need to indulge your power and macho strength. I was committing fratricide by running riot with the other dogs to drive out, chase and savagely attack wild animals of every sort. All those animals ceased to be animals, brothers and creatures and became "wild animals" and "prey" just as you consider them to be. What is it like to organise a hunt or a war, to strive for perfection of this continuous and absurd violence?

My wary friend, the cat, is right when he says that we never see "beware of the cat" written on any gates. Only the misguided "beware of the dog". This is your doing, to defend your property, your home and your life. And to think... I used to be proud of it. We never rebelled against all of this. We only have our instinct and not reason or logic. We don't feel hatred. We don't even hate the other animals that hunger has driven us to kill. You, on the other hand, often need to feel hatred before you can kill. And yet, although I was sewn up in a rough and ready way, I am happier now than I was before to be so close to the human race, both the good and the bad among you. If my blind man doesn't hate anyone, how could I?

Do me a favour, human. Show me a real sign of your humanity, intelligence and goodness. Try not to humiliate me. I am hurt more by words than sticks and stones. Stop "domesticating" dogs, cats and birds for personal use and abuse and as a source of income from your degrading animal "shows" in circuses and other cruel places. Don't turn us into "abortions" and grotesque caricatures of yourself. We may appear to be less like animals, but you appear to be less human.

There is another insult that I cannot tolerate: dog shows, those canine equivalents of your beauty pageants where your most attractive, and often least intelligent, young ladies are paraded. I can already hear the breeders and judges using bad language. I can hear the squeals coming from the owners of "toy" dogs. But I cannot keep silent.

Human, don't imagine for one minute that you can become like God when you tamper with the genetic code and the mystery of life itself. And because you are not God, fortunately the only beings that you can create in your own image are your own children. Don't add insult of God to the injury of other creatures, be they your children or your pets.

We animals are different from you. We like you and we like ourselves for this very reason. Aren't you satisfied with having us as friends, just the way we are?

You may be immortal, but you are not divine and, even if you have the nerve to turn us into carbon copies of yourself, just think how boring it would be. I am well aware that your species will never give up trying to attack the human race, just as Cain attacked Abel.

Dear human, my incorrigible friend, at least take your hands off the dog.

THE CAT

Dear Humans,

lthough I keep my distance with due respect, here I am, next to you as always. No need to get excited if you don't see or hear me straight away. I can't stand excitement. You ought to know. You are always going on about how cool, calm and collected I am, and this goes to show that you agree with my attitude and like me because of it.

It's true. I don't interrupt, shout or make a racket. I can dart off and slip between the most valuable display of glass without breaking a single object. What's more, when I want to find out what's going on in the house or outside, I don't have to pierce your ear-drums with earth-shattering barks (as the dog always does).

Although I am openly speaking my mind, I know that it's not very nice of me to attack the dog in this way. But I also know that deep down you are just like me because, like me, you enjoy complete freedom.

Very few animals have my freedom, especially the so-called pet. If I don't like my master or anything else for that matter, I behave in a dignified manner, never giving in and never making allowances even if I am hungry. I certainly wouldn't drool (like a dog) for you to stroke me (yes, I am repeating the comparison). I do miaow for my food, to accept your left-overs or perhaps taste the latest range created especially for us cats and dogs by the wizardry of animal dietetics. However, if I haven't eaten for days, I miaow more softly, not more loudly, and this is usually enough to make you understand that I am at the end of my strength. I want your affection, friendship and your strokes for free, without having to give something in return. If this is not the case, I move house and change masters, even if this means losing the person who thought he was my friend.

Someone once said that I must have descended from the Pharaohs of Egypt because I am so free, beautiful and dignified. We have the same self-possession and dignity with just a hint of mystery. Of course, this lineage is not possible because the cat was roaming around the world and through history before the time of the Pharaohs. In fact, it was they who adopted me; I never chose them. Where do I come from, then? Never mind! Let's keep the mystery. If we wanted to be paradoxical, we could say that the Pharaohs descended from the cat. Just as you, perhaps, descended from the monkey.

Perhaps this explains the story of our descent: at the beginning of our time, a crazy cat became restless so he busied himself by cleverly imitating the pomposity of the Pharaohs as they are depicted in paintings and monuments. Then, some alert artist captured that mimicry and it became a sign of high lineage. Even if it is a tall story, it's a lovely story and does us great honour – on condition, however, that it stays a story. Believe me, no sensible cat would swallow it. As most, we find it amusing. There are perhaps times when we try to live the part, but we are well aware that our fascinating stillness, self-control (in truth, a mixture of laziness and wisdom), the changing – sometimes odd – colour of our eyes and the variety of our fur is not enough to make it a convincing story.

Luckily, the truth is quite different.

It is much simpler and far more beautiful. I hope you find it to your liking too. You have to remember that the truth of the matter is that all cats are either looking for someone or living with someone. Once they have found their person, they know that they also have to adopt that person's family and friends. For this reason, no self-respecting cat has ever wanted nor would ever want to be a human being. Only a cat with a screw loose would think that he had to become human to dominate the human race.

You love us so you are well aware that the cat is never a slave to the human. We are your friends and companions through thick and thin but never, for any reason, are we slaves. We have the use of your homes, your property and take up your time. These things are more necessary to us than oxygen. Unless there is some other reason for conflict between us, we never take advantage of these things. We are content with you, in your home, your own domestic environment which you share with us. We both know our place and there is no need for even the shadow of one of those worrying role reversals which only your psychoanalysts can explain (but never, believe me, with any positive result).

We are content to live close to you, to love each other, to procreate and multiply, to be loved, to purr and enjoy waiting for your return. We are sad when you go out but we can use your absence to explore the neighbourhood, wheedle a bit of food from others, go courting with cats we haven't met before. We never stick our noses into your business and never deserve to be shouted at or kicked for having meddled or whimpered or pestered.

And yet...

And yet, if you think you know all there is to know about me, you are much mistaken. You will never know all there is to know. I will always be a bit of a mystery to you and you will never be able to fully understand me. I believe that the constant desire to discover something new adds something special to love – even "human" love.

I don't like preconceived ideas. I'm not in favour of the saying "man's best friend". And thank goodness that "the human is the cat's friend" has not become another hackneyed phrase. These labels are meant for the dog, and you and the dog are welcome to them.

Unless I'm dying of hunger, I don't accept the first person, the first master who comes my way, nor do I necessarily accept the house where I was born or where they move me. I always look around first. I may be extremely happy in the house where I was born, or I may never want to set foot in it. It depends on how I feel.

You could say that I am emotionally independent and absolutely free to "marry" you and "divorce" you. After all, can love be true love if it isn't free? Whether your intentions are good or not, you often measure love against the degree of slavery into which you men, women, parents, children and friends have fallen. We cats don't do this. The measure we use for love and freedom is completely different from yours. This is the reason why I almost always get along with you and stay with you for life.

Sometimes I am envious and jealous of your "friendship" with dogs. I often watch you getting on together, showing your love by drooling over each other as though you could face anything as long as you are together. Then, when I see you uncontrollably cry your undignified tears in unison, I pull myself together and have to snigger – no offence meant. At times I also try to give or receive comfort, but I never go over the top. I might purr a bit, rub myself against your legs and occasionally jump onto your lap, but I never go too far. You may say that this means I am not as happy as the dog. Not true.

I am just as happy without even wanting to know the reason why.

In a nutshell, I don't want to live the feline equivalent of the "dog's life" I often see lived by both dogs and people with all its traumas and incredible frustrations.

The cat is made differently. I am the product – the victim, if you like – of much longer research, better thought-out control and more demanding mistrust. Even if all this is useless and often fails.

This is what my instinct and affection strive for. I express it in code, in my own language: that of the cat, not the dog. You mustn't demand love, don't curse and betray if you don't have it or you receive it differently from the way you hoped. Don't be offended if I tell you that, as friends, we have to be equal in everything, in freedom and dignity. There are risks involved for both of us. I could lose you or you could lose me. But once we have established our relationship, losing one another is virtually impossible. Your newspaper reports state this quite clearly. No matter how great the distance, hundreds of miles perhaps, separating cats and dogs from their homes, they almost always find their way back. I will never lose your scent or that of your home. I can always find my way back to you, your home and your troubles. However, if you lose me, all you can do is offer a reward for whoever finds me.

If we were really free we would never lose each other. I like living in your home just the way it is, and with you just the way you are. But how many cats have to pay an extortionate, painful and humiliating price for their freedom every day? They have to put up with lack of affection, blows, starvation, abandonment or being kicked out. Often it's simply because they haven't learned to be the "pet" a neurotic and hysterical mistress wanted or some cruel children expected.

They all say that no-one loves his master's house like

a dog. It's true. I don't love your house like he does, but I love it as much as he does. Nevertheless, even if society is near to becoming a lager hiding behind an attractive facade, I have never seen "beware of the cat" written anywhere.

If I had to, we all know that I could scratch out the eyes of a thief or murderer. But I hate conflict and never look for trouble. In fact I run away from it, at the risk of being branded as a coward and selfish in the storybooks. I never enter into a contract or sentimental pact with you. We have a tacit agreement. I only have to open one eye while I'm snoozing to see whether you are content or unhappy, and I never deny you a few purrs.

You always talk lovingly to your dog. The most you ever say about me is that I have become attached to you. However, if your words have the same meaning for animals as they do for humans, then I too love you. In fact, when I love you, I love everything about you, just the way you are, men and women, good and bad, the selfish and the sensitive. I love the poor, frustrated ladies living alone in vast cities, looking after masses of stray cats. When they don't squeeze me too hard, I love your children. I love your silence. I love being able to live with you without being a nuisance. I can be good company without overdoing it. I am there, breathing, purring. My life is better because you are there and I can feel your presence. When I get fed up or I'm feeling a bit irritable, I leave you alone and don't complain. Dear humans, we are company for each other. Just as many cats have always been good company, not just for many children but for the saints as well. I often think about Clare of Assisi's cat, her only companion for many years when she was stricken with arthritis. Above all, however, I think about my master. Whether he be a saint or a scoundrel, I either have to love him the way he is or not love him at all. I hope that the day will come when our relationship is not dependent upon the need for food or affection – that is to say, necessity – but upon the freedom to choose friendship.

THE ERMINE

Dear Humans,

I am an ermine, a mother with many babies. Unlike my children, I happen to be still alive. Fortunately, or unfortunately, I have survived while they have become valuable fur coats belonging to sovereigns, princes, popes and the wives of some of the wealthiest men on earth.

My life is hopeless, and at times I regret my instinctive drive to bear yet more children every season, as though I am committing a crime. I know there is no way out for me or my race. Our beauty is our death sentence. You rob us of our beauty to create a luxury market. This theft feeds the pride and exhibitionist tendencies of individuals and the selfish, privileged classes.

We have become an historic symbol of the power which makes money synonymous with control, wealth with the ability to indulge every whim. The first to do this were kings and princes with their respective women, patriarchs and judges, all influential laymen and clergymen. The ermine continues to this day to be one of the surest standards upon which to judge the wealth of men, women and families. As usual, it was women who started the greatest massacre. Even if the powerful ceased to use our fur as their symbol, feminine ambition would be enough in itself to guarantee both intensive ermine breeding and their capture and slaughter until the end of time.

Our future holds nothing for us except brutality and violence. Of course the hunt is different nowadays because of the more selective and cunning means to capture and kill us without damaging our precious fur. Although it lacks the most basic common sense, I expect that "ecological hunts" will be next, specifically designed to have our fur intact. You need us all to be safe and sound at the right moment, and can't just kill us indiscriminately. By doing this, paradoxically you become the most rigorous defenders of our species, guaranteeing the continuation of the ermine whatever the cost. As worshippers of the "ephemeral" and the "image", you passionately fight not to be without our beauty and are prepared to pay "astronomical" prices, to use your word. Those few of you who can afford it must never be without our splendour in your palaces and high society salons.

We never hear any really hopeful, concrete ecological news. Anything we may hear is immediately denied. Some time ago I was relieved to hear the news that many women are now wearing pig-skin. I couldn't say whether this means actual fur coats, or skins from which jackets, gloves and other whimsical fashion nasties can be made. You are capable of miracles even from a pig-skin. In any case, we ermines are not consoled by this. In fact, I believe that this fanciful new idea simply means that anything goes when it comes to that great, devouring idol called Fashion, whether it be ermine fur or pig-skin. All it means is that things are getting worse for us and for our brothers, the pigs. Men give in to women's whims and there is no end in sight. In one way or another we animals are literally risking our lives continuously.

However, I do not want to appear to be a boring and annoyed moralist. If you weren't such cruel, war-like consumers, I might even be pleased about the fact that you value our beauty and enjoy the taste of animals like the pig. According to the first of your holy books, Genesis, you have been given the go ahead to feed off all of us, without any exception for any reason. But I definitely know that you weren't told you could be cruel, or capricious enough to enjoy seeing us die.

You already have all that we have to give. What more could you possibly want? For example, nothing is wasted from the pig as you know. Before the advent of nylon, even his bristles were valued. You may or may not know that these days the pig gets by the best he can. If you challenge him he'll become as fierce as his close relative the boar. I have been told what happens in Italy to herds of pigs that become ferocious: the mafia uses them to punish people for their mistakes. The pigs eat them alive. Imagine that?

No, our brother the pig is not well-read, to use your expression. He is praised only for his products: ham, chops, bacon and sausages, liver, black pudding and trotters. Unfortunately some of the fault lies with your Gospel, your book of life. Pigs are far worse off in this book than any other. Even Jesus, when ridding a poor man of the demons that had inhabited and tortured him for a long time, grants the demons' wish to be transferred into a herd of pigs. Jesus grants their wish and the horror-stricken pigs, who were grazing in the pasture, plunge head-long into the lake below. This sight appalled the owners of the pigs who had now lost their capital investment. Showing no concern for the man whose life must surely have been worth this loss, they begged Jesus to go and perform his expensive miracles elsewhere. I can't see that there is any mystery in this story.

Let's get back to our discussion about ermines.

Humans, if you carry on your same cruel practices to have our fur, you are bound to fail sooner or later no matter how much your women want it. We have never been proud of our fur because it has always reinforced the power of earthly riches. We have never been happy to provide vain, social-climbing women with a status symbol.

We represent beauty, not power. The two must be kept separate. You are the ones – with your beauty treatments and money, your wars and peace-time exploitation – who have always confounded beauty with abuse of power, beauty with wealth, artificial and natural beauty. It's sad for a regal and solitary animal like myself to think that our beauty so rarely represents goodness, justice, real authority, celebration and the prestige of creatures who live for other creatures even if they don't dress in the same precious way.

The dynasties, the ermine-clad sovereigns, the old and powerful clergy who at times have been in conflict with the world's rulers: these are all declining institutions and people in a state of crisis. Either for the sake of love or because they have no choice, the "look" is changing for all of them. Why continue this power and its immaculate symbol (alibi or true inspiration?) by transferring it onto women?

A person who idolizes power is an imprudent and blind idolater. However, a person who uses an ermine fur or a whole collection of furs to turn an ugly woman into a less ugly woman, or a beautiful woman into a more beautiful woman is a forger and an enemy of common sense despite his allegiance with good taste. Who can survive the theatricality of high fashion where the only winners are the changing whims? There are very few great masters of frivolity with the morality to recognise and really appreciate the pureness and innocence of our fur, our clothes, this symbol and trophy of odious massacre in your violent hands.

With anguish and compassion rather than the anger

you deserve, many species of animals are gauging the rate of your moral collapse, watching your humanity disappear even more quickly than your common sense.

You are also a species on its way to extinction. The reason is not so much because of a biological deficit but because of over-spending on your inventive powers, a flaw in your intelligence capable of reeking havoc.

When you are hunting us down, when you feel like a king in the conquest of a gift for your queens, you don't realise that those pure, precious symbols will almost always dress violent or potentially violent people.

You are totally naked with only your presumptuousness to cover you, and I fear that you are fast approaching your own requiem.

Dear humans, I beg you to help me not to give up all hope for you.

THE SWALLOW

Dear Humans,

I am the swallow, whom you and your poets love so well. The naïve old lines: "swallow, wanderer so small / always singing the same song / as you perch upon the wall / all morning long" have always given me the impression that my twittering black and white beauty was enough in itself to please you.

I am glad about this and am sincerely sorry to say that I have never had quite the same liking for you. As far as I can see, over the centuries you have never really learnt anything about me. Few people have come to know me well, and few know what my real purpose is. For you humans I am the one who announces the arrival of Spring, the ornithological equivalent of the primula. According to an old proverb, if I am "under the roof" on the 21st March, then Spring is here. But how can you still believe in these old sayings? And as for the seasons... everything is upside-down! It's cold when it's supposed to be warm, cool has become freezing and the winds are unpredictable. We are beginning to lose our hitherto infallible intuition for the seasonal cycles which, until fifty years ago, was more reliable than your thermometers, clocks and calendars.

You get upset if we don't arrive on time, but then, having realised that the 21st March has come and gone and it's still freezing cold, you question whether it really is Spring. You are even disappointed when our absence tells you that Spring is still a long way off. But it's no use trying to give you a sign with our absence. In some countries you have realised too late that we aren't coming back and, later still, you have asked yourselves why.

The truth is that our work has been compromised and ruined by you humans. Until recently we cleaned the air during the day just as our brothers, the bats, cleaned the night air. These were wonderful and enjoyable work shifts. Everyone did his job happily and we were all friends. Now much has changed for the worse. Even in the places we are still able to reach, the shifts have been reorganised and we have to make do with trying to do the best we can. We leave at sunset and are replaced by bats, soaring sideways into the soft evening. With the first rays of the morning sun, we whiz about and twitter once more. We quickly and happily remove whatever impurity has been left by the bats and probably seem a little crazy with our chirping and chattering – which you find either cheerful or annoying.

The problem is that even though we have everything we need to accomplish our task – infallible vision and speed as well as intuition – we now also need a stomach able to digest poisonous insects which, either by mistake or because they are like kamikaze pilots, fall into our small but deadly red beaks. Because you have poisoned both the air and the insects with your smoke and smog, we are also gradually being poisoned. The insects do their job and we do ours but we are both heading for extinction if we don't solve the problem. Very soon it will be impossible for any swallow or insect to purify the air, the earth, vegetation or water. We will even be afraid to eat the already contaminated earthworms which have always been our very favourite food.

I know that swallows are beautiful. However, dear humans, I don't say this out of vanity. It is a simple and very pleasing fact. Black and white with our ash-grey collars, we are much more elegant than the likeable yet foolish penguin. I am also more elegant than you when you dress in "tails", de rigueur for your formal, upper middle-class occasions. My white chest is far more elegant than your best silk dress- shirt and my tail is finer than the finest velvet.

Yet all this beauty is in peril and condemned to death if you don't come to your senses. If I linger in the big cities, I know that my shiny black colour will become grey and opaque. The grey is not the colour grey should be, but a dirty, indelible, lasting grey which has no right to be called a colour. And don't tell me that because I say I'm beautiful, I am vain and this is a figment of my imagination. That grey is not even a temporary coat of colour. If it was, all I would need is a quick bath during my arrival or departing flights to recover my shining feathers. Not even the most concentrated (and most poisonous) detergent will do the trick. You are even destroying colours!

What is more (no, don't start showing signs of irritability), you can't tell me that this complaint isn't worthy of the "queen of birds". It's not a complaint. It is a warning. I am trying to keep a flicker of hope alive. It is my civic duty to warn you about the dangers and urgency, and it is the duty of you humans to heed this warning.

As I was saying, the insects which have always been precious to the imponderable and very delicate balance of the air and the environment, to the greater scheme of life and nature, are disappearing. Instead, swarms of almost unidentifiable larva are increasing, born out of all the visible or invisible active poisons that exist everywhere.

"As free as a bird" used to be the saying. I was the symbol of freedom. I could fly away and still be domesticated, with no restrictions and no conditions. Can I still make this claim? Without lying to myself, pretending otherwise to myself or to you? I am tired, worn out and disappointed. We have had to make very difficult yet necessary choices: to leave the places and countries where we can no longer honour our obligations with any good result, where our very identity has been altered by your consumer culture and your moral and environmental decline.

Many of our flocks have already left for good, and most of you haven't even noticed. Only the old people among you and a few derided ecological movements are aware of these alarming changes. Our twittering once wished you good-day in the morning and good-evening at sunset. But who notices our almost total silence these days? Dear friend, as the years go by, you are depriving yourselves of beauty and losing more and more friends. We are also suffering as a result of the loss of our friendship.

I must confess though, I am not totally pessimistic. I believe that our escape, our disappearance could become a brotherly challenge for you. I hope that you are aware of how solitary your existence has become in the current vast "desert" of your environment.

We will go elsewhere and feel sure of finding new living space, different air, other insects, sun and purer – or less impure – water. We will make other friends who will be inspired by those feelings that we used to inspire in you: happiness, hope, joy. We will be happy with them. Who knows… if you come to your senses, perhaps our children, grandchildren and great grandchildren will be able to renew the marvellous relationship we remember having in happier times.

This undamaged paradise must surely exist elsewhere. We will find it at any cost. Despite the bad times, we want to find the Adam who has remained within you all.

If necessary, we want to help a new dove return from a new ark of salvation from the "nuclear flood" to tell you that the catastrophe has been avoided.

My friend, no matter what the season, I the swallow, or perhaps one of my descendants, will return to tell you that the flood has abated. Don't despair. If you won't save us, then we will have to save you.

THE SEAL

Dear Humans,

y some miracle, this old seal has avoided customary massacre by your hunters seeking my skin and fat. When I think of what's in store for me, and the future of every other seal on earth, I think in terms of a greedy, violent army. The few people who respect and fight for our survival are precious to us. They are courageous people from all classes of society, politicians, scholars, famous actors and actresses and sincere, enthusiastic ecologists. However, I am still pessimistic. At this rate, no-one will manage to save us.

In the so-called animal kingdom, we seals are neither cynical nor romantic. We have always known that our fate was to be prey for human greed. Even though we don't make a fuss about it any more, the pain has not diminished. It also has an indirect effect on you humans because, the more animals you destroy and wipe out, the more you prepare your own disappearance from the face of the earth.

This is my only reason for writing to you. I hope your conscience will be pricked by the work of those people I mentioned before. I also hope that you will have a sensible and effective plan for co-ordinating your efforts to guarantee not only our survival but also that of other species of animals. Sensitive politicians, intelligent artists, well-informed ecologists, scholars and clergymen are all capable of reviewing life on our planet. These people have the ability to save us from extinction. Although we may appear to be clumsy, we are very sensitive and one of the most deep-seated and sincere feelings we have is gratitude. At this rate, we are all heading towards the end of the line. However, the trip has not been made in the company of good friends.

Just before I go to sleep I try my hardest to understand what the heroes of your news reports and history are like. Trying not to lose hope, I enjoy wondering which one of them should be given the task of protecting and defending our various species.

For example, I would willingly like to see one of your principal theatre and film actors, with their resonant voices and regal stature and bearing, defending the lion. Although the so-called king of the animals, this species is perhaps under even greater threat than we are. I would like to see another of your well-known actresses, known as a tigress, actually defending tigers (despite the cliché about tigers being able to defend themselves). I would give one of your film directors the task of prohibiting the generation of bastardised breeds of animals in the circus, horses in particular. Horses are pure, strong, almost divine beauties of creation, as dynamic and powerful as they are intelligent. A great director always knows how to handle the finest actors, even blonde bombshells as you call the silly women he either chooses or are chosen for him to test his Pigmalion ability to turn them into magnificent butterflies. I would give a multitude of cheerful and very elegant penguins to an old scholar. They would be excellent company for the enlightened and liberal leader of a people. If they want to, all these famous humans, to mention but a few (but as good as any other), can help us a great deal. In return we could help them become better known (however, let's not mention advertising, as this has become one of the most ambiguous words you have invented). When you read about it in the papers or discuss it during a delicious meal of roast veal or fried fish, doesn't the idea of looking after us animals warm the cockles of your heart? While greedily stuffing themselves with hare and pheasant, even hunters have the courage to look each other in the eye and talk about "contained and selective hunting". From the serious nature of their discussions I would say that they actually believe in what they're saying. But the conversation does lose a bit of credibility with the umpteenth helping of jugged hare....

But let's get back to us seals.

Humans have always known how generous we are. We give you our skin and our fat. And no human has ever thanked us. With their raids and increasingly more scientific and deadly shooting parties, arctic poachers are gradually thinning out our numbers, the large families which make up and maintain the balance of polar fauna living on the glaciers. But for you humans, what difference does it make if the great wealth of animal life on earth is being violently reduced? The most savage predators are not afraid of anything. They don't even take our growth and reproductive cycles into consideration. As they plunder and murder, their only interest is financial gain. They watch as we clumsily try in our terror to waddle to safety, but they only see handbags, jackets, shoes and vats of valuable oil. We are just a commodity, a product like any other, nothing more.

Human, your vision is distorted. You are so shortsighted. You become greater or smaller in line with your earnings and financial interests. All you can see and calculate is your immediate profit. You can't even see that with this cruel and insane greed you are going to reduce your stock of merchandise until there's nothing left.

Like the whale, we are also the victims of literary bad taste. Even if based on a worthwhile idea, the stories grow mainly around the person who challenges, abducts and destroys the animal and not around the animal who is the preordained victim. The hero is always the fearless human who tirelessly takes all manner of risks to butcher the "enemy" – and that is to say, us. And if this human happens to have a leg missing and manages to defeat (he is, however, also defeated) a mountain of enemy flesh like a whale, the result is a literary work of art, a sea-faring adventure story about the most famous white whale, *Moby Dick.* What's more, if the book is turned into a film, the cinema or television audience will sympathise with the surly Captain Ahab and not the whale. The Bible also talks about a strange, immense whale and of a frightened man who ends up in its belly, the prophet Jonah. The book is a small masterpiece of humour. But your interest is in Jonah, not in the whale who accommodates him in his dark abdomen. If any fisherman should see an island of meat emerge from the arctic waters, no matter what the cost all he can think about is its capture. Occasionally we come across specialist hunters, with some remnant of conscience, who are at least able to judge whether the whale is pregnant and should be left to give birth first. Not even a shark's savagery could avenge your many massacres. How can the violent death of a human devoured by a shark, or a few dozen rash fishermen mutilated or killed by sharks, that are only acting out of self-defence and never for vengeance, be considered as repayment for all the exterminated seals and whales?

Humans, we are about to disappear but so will you unless we can find a way to avoid it. You are the rulers of the earth, the sea and the sky (but, fortunately, not of "heaven" as this is not made of "sky" as you well know) and yet, you will be the worst off. As we disappear, like the whale and other fish, your destructive zeal will grow with each indiscriminate massacre, like the "antiphon" before a "psalm" about the end of humanity, or at least the end of this generation blinded by your egotistical well-being and wealth at the animals' expense.

The sums are easy. All we need is ten, perhaps a thousand of your oil tankers – true floating whale-like death traps – to be hit, accidentally or on purpose, by enemy planes or gun-boats or some other product of nuclear warfare. When their tanks of petrol are emptied into the open sea, the result will be death and disappearance of marine life, sterility of the oceans and seas and pollution of the coastlines for years. Surely you will at least be upset if you can't eat your precious lobsters and all the other top quality fish you've grown to enjoy? The river-like oil slick which seems to keep you all happily and irresponsibly afloat, be it on the sea, the ground or in the air, will also cause you to perish and die. You will have an agonising death, lasting about ten years, a century at most, before you become extinct. Very few people live to be a hundred, so why should you care? You are happy to sit at your table, licking your lips and savouring the taste of fish like "whitebait" which could have grown to weigh a ton or two instead of meeting an untimely end on your merciless plate.

What more can I say, dear humans?

Despite everything, I have faith in what is left of your guilty conscience and your common sense. I will ask just one question. Although you are greedy and irresponsible, won't you at least try to respect the "whitebait" and the eggs of dozens of other species of animals, to avoid the useless and mean "slaughter of the innocent" in the animal kingdom? Aren't you content with already being specialists in the slaughter of innocent children with the spread of injustice, hunger and war? Can it be that you only want Herod as your patron saint?

THE MOUSE

Dear Humans,

I am a mouse. Please don't scream and run away, or try to find something to hit me with or chase me off.

I know I give you the creeps and make you cross. Especially you, dear lady, and all women in general. I'd just like a minute of your time to tell you something that concerns us both. I'd like you to understand that, despite everything, at times I am the animal you most need for your health. In fact, this is the one thing that you all seem to forget.

I am not trying to cheat. I am just saying that many scientists and doctors don't seem to want to mention this fact or, at most, they only discuss it amongst themselves and never with any conviction. I am the "guinea-pig" par excellence, the "crème de la crème" of guinea-pigs. Everything that you people suffer from, I suffer from first. The more unknown and terrible the ailment which afflicts you when the first symptoms appear, the more I also suffer in your laboratories where you prick me like a pin-cushion and slice me up as though I were a loaf of bread. I pay with my life so that science can develop more defensive means to treat you and save your lives.

These lies and tricks have been going on for centuries. These days, some of your more honest scientists and doctors claim that all research and experiments on known and new infections can, if necessary, be carried out on a few cells and tiny samples of your tissue and ours. But hardly anyone has been able to show this convincingly. I know that, from the very beginning of human life, the price for all scientific progress has always been war and the cruel practice of vivisection. In short, a life must be lost or made to suffer in order to save other lives. The greatest insult is that you continue with this platitude, pretending to be unaware of the one thing that would do you honour: to find what you are looking for without sacrificing us, just as dogs have also uselessly become innocent victims.

I certainly don't have any say in the matter, nor can I claim any rights. I cannot defend myself by squeaking. Who has ever paid attention or listened to the "roars" of a mouse? No matter how deep a breath I take, I have never managed more than a sharp squeal, an anguished snap and usually no more than a faint hiss. Who would listen to such a protest against pain?

So be it... but I want to tell you, the ordinary human, not the doctor or the scientist, something about the imminent dangers despite the continual pharmaceutical and scientific sacrifice of guinea-pigs, white mice and other animals. You would be far better off than you are now without us, our sacrifice and our compulsory, unjust contribution. You would be able to detect the lies constantly told by the despots of vivisection experiments. Either out of pride or laziness, they are telling you a load of nonsense.

I want you to know that even without my sacrifice, without all the open or secret massacres of guinea-pigs, you would have been able to achieve scientifically, the means for better medical prescriptions and the ability to prevent and cure many ills.

Why have we always had to pay for your security? You people hunt us and eat us (not us mice, I'm talking on behalf of other animals); you destroy us and poison us. Hardly anyone, even the less cruel among you, ever thinks of offering up his own life for his brothers in the human race, for the sublime glory of conquering human suffering on all fronts (and animal suffering, as a veterinary friend of mine says). In recent years, I have often heard it said that you have obtained some good results with cancer research even if an actual cure has not yet been found. If we take pride in this it's because we have been of some use in this struggle. Our guinea-pigs, inoculated with cancerous germs, have given you some hope even if the results vary from guinea-pig to guinea-pig.

I will explain myself better and give you proof. Some time ago, while gnawing at some old files in an archive occasionally visited by a group of medical scholars and researchers, I almost destroyed some valuable scientific information favouring us. If you should still feel sensitive about these things, the result will shock you. So don't take offence and have the courage to understand what I am about to say so that you can destroy your collection of mouse-killing devices stored in your cupboards. That document, read but badly received at an international medical congress, literally says things that you really ought to know about. A specialist in the field (I won't mention his name so that his top-ranking colleagues don't try to have him struck off) says: "Can anyone be truly pleased with medical progress today? How many people know that 60 per cent of existing diseases are iatrogenous, that is to say contracted from drugs? And that medicine causes 6 per cent of fatal illness and 60 per cent of malformation? And 88 per cent of still births? And that 15 per cent of people admitted to hospital leave with another illness, often without having been cured of the previous illness?"

I know, my friend, you always bring up the thalidomide business. It's true that this drug, with its funeral procession of phocomelic children, is perhaps the most sensational example of the results of medical research stumbling in the dark. Thalidomide is not alone, however. It is in good company. I'm sure you remember "clonquinol", the anti-diarrhoea drug which caused Subachute-Myelo-Optic- Neuropathy in around thirty thousand Japanese who became irreversibly paralysed and blind. Of course it had been experimented on animals and declared free from side-effects. What about Silbestrone, the artificial hormone, which caused vaginal tumours in the daughters of women who had used it during pregnancy?

All right, my friend, I have only mentioned the most deadly like thalidomide, but would you like to know the reason for so much slaughter caused by these medicines? The only reason is because, after being experimented on mice and guinea-pigs, they are declared harmless, in fact miracle cures for men and women. Thalidomide itself was marketed after it had been proved non-harmful in many species of animals, the rabbit included. After the catastrophe new experiments were carried out and only then was it discovered that it was teratogenous in the New Zealand white rabbit, one of the existing one hundred and fifty species of rabbit! Many vivisectionists admitted: "The catastrophe would have been avoided if we had experimented and tested it earlier". They are obviously lying, telling blatant lies...

No, no, my friend, I haven't finished yet. Listen to me for a few more minutes. Overcome your trembling fear, your nausea and your instinctive desire to run away. There's more. If all medicines which are harmful to just one of the many sectors into which animal families are divided were thrown away, there wouldn't be a single drug left.

Not even aspirin, which is poison for the cat; nor amyl nitrate which raises the pressure in the human eye; nor digitalin which raises blood pressure in dogs and mice. And above all, at a time of dire emergency, penicillin would not have existed to save the lives of millions of humans, but to systematically kill the guinea-pigs. As luck would have it, in the thirties and forties when it was quickly decided to use penicillin, there were no guinea-pigs in the experiment laboratories – only mice. Had the antibiotic been injected into a guinea-pig, it would most certainly never have been authorised.

You can, therefore, thank us for avoiding a defeat, an epoch-making catastrophe. Try to remember this when we give you the creeps and you plan our extermination like a Nazi specialising in pogroms with all the stupidity of an illiterate.

Why can't we be friends?

We all know that if you weren't so selfish, lazy and dirty, polluting the cities, your homes, your cupboards and everywhere else, either indoors or outdoors, you wouldn't receive so many visits from us which, most certainly, can sometimes cause infection and contagious disease. We could live together – at a distance – with much greater mutual respect. It's true, we spread dirt and poison, but only because it's there waiting for us, mountains of it prepared by you. We only carry the plague when you have been the first to pollute your own surroundings. In myself I am a clean animal, and if you were to look at me without prejudice and disgust you would see that I am also beautiful.

I have to admit, I often dream of living with you in friendship and with affection. Not, however, in the way we do now like a few token slaves cooped up in cages where you keep hamsters and guinea-pigs. But free and well-fed like they are, not scavenging for food only to have to flee from your raging temper. Didn't you know, my friend, that we were in Noah's Ark too? We respected each other then.

You may not believe it my friend, but I love you and,

like you, I am one of God's creatures. To prove it, I have dedicated a little prayer to you, written by a mouse friend of mine who loves poetry and uses a very human-sounding pseudonym.

Lord,
hear my heartbeat:
they have set traps
everywhere, just for me!
They want to capture me,
they want to catch me,
not alive, but dead!...
Lord,
raise them
from the dead...
Lord,
why do children run away in fear
when they see me?
I want to make friends with them
but they won't wait for me.
Lord,
don't let me run away
from things I don't understand...

THE STORK

Dear Humans,

I am a stork. My life lies somewhere between fantasy and reality. A world which only respects imagery and symbols, but lacking the innocence of days gone by, has turned me into a flying fertility symbol and birth announcement. I have always lived and built my nests on the roofs of your tallest houses, cathedrals and ancient towers. I have always enjoyed living with you people and am often touched, often amused and at times horrified by the sight of your slow-witted approach to each other. As I raised my own young on the roof-tops, I too rejoiced every time a baby was born.

Perhaps you haven't noticed, but for some time now I have gone to live elsewhere. Elsewhere from where? From wherever the smoke, noise, intense pollution and thoughtless town-planning has engulfed a city. It reached a point where we animals were under siege, and so I looked for shelter in small towns where your unruly, concentrated urbanisation could not endanger the future of us storks as it is doing to you.

Now we are beginning to return among you adult and infant city animals. Being intuitive and observant, I realise that you have begun to understand how protecting the environment means protecting your very lives. I therefore decided that taking a few risks in order to give you a hand to overcome this difficult struggle was worthwhile. This help does not consist of once more boosting the sale of pretty announcements proclaiming "joy" over the "happy event" when one of your children is born. My picture, holding a pink or blue bundle in my beak, may be very sweet but it has never been more than a parable on a greeting card. I am actually very serious about living together with you. If we once had a duty and the ability to inhabit peacefully the same houses, then surely, with good will on both sides, we must be able to do it again and this must mean happiness for both of us.

So here I am again. For now it is a trial run. Perhaps you could say that I am exploring your river. Although a beautiful river, it has been scorned and degraded by the input of toxic waste. I feel like a bride or a new mum who, together with her husband, goes to look for a suitable little one-room flat which might hopefully be pretty as well as comfortable. I don't expect a great deal but it has to be suitable. Of course, I prefer towers, houses and mansion blocks or huts along tributary rivers where great factories do not discharge their foaming barrel-loads so fatal to the fish, birds and humans. But I want to win the bet that I can return and choose to live by the river. I don't know if you remember one of the reasons why, until recently, I had disappeared from the river areas as well as the smoking city chimney-stacks. It was because of the boating craze. People were trying to navigate the rivers in flashy high tonnage yachts and boats not much smaller than those paraded along the fashionable coastal resorts. Noxious fumes had reached danger level. Then, luckily, the younger generation began to listen to reason and realised that it is much better, healthier and more sporting to go canoeing on the river, using elbow grease instead of fuel. During my early reconnaissance trips before deciding to return, I saw crowds of hot and sweaty yet happy young people travelling the river in canoes. I understood them – I, who have always admired country and city from the sky

– because seeing a marvellous country from the river rather than the river bank is a completely new experience.

It's an uphill struggle, but I believe that we are winning in our attempt to overcome once and for all the bewildering and perverse mad manners (another form of pollution) associated with so-called tourists. Perhaps unwittingly but none the less dangerously, this behaviour has affected young people of all origins and all classes of society. I once flew behind an overcrowded group of canoes being paddled by some young people. There were three boats, crammed full of shouting, smoking youngsters. Panting, red-faced and sloppily dressed, they happily reached their destination while I, wheezing and feeling bitter, was at the end of my strength. It had been more of a river battle, being bombarded with fruit peelings and various left-overs, than a sporting, ecological mini regatta. Recently however, I frequently see organised river trips made with a precise ecological aim: to clean the river rather than dirty it. This is another reason why I have returned. Those who behaved like bulls in a china shop, leaving a trail of debris behind them, are being replaced by people who hardly litter or pollute at all. I probably have a Manichean philosophy and will pay the price of my positive attitude with my optimism. It doesn't matter. Changes are being made. I give my blessing and want to see more.

When I first went north, where it's much colder, I thought it was cleaner with more respectful people. I lived in a variety of spires and safe chimney-tops crowning those neat houses made of stone and gaily painted wood. But even there, where nothing even remotely resembles an industrial city, it was the same old story. The only difference is that further south pollution increases in an "untidy" manner, whereas there it's "tidy".

In the north, our baby storks run the same risk of hatching already poisoned and handicapped as they do further south.

In all modesty I can say that I feel I have the same vocation as the dove on Noah's Ark: an explorer who compares floods in various European countries. Rather than carrying a bundle containing a baby, I have to find out where I can take the olive branch, be it real or symbolic, which I carry in my beak.

I have returned to offer this olive branch to you. More than ever before, my friend, with all my heart I would like to become an emblem once again, the omen of your vital desire to overcome all the poison and environmental outrage and to guarantee a future for all creatures and all families. At times I am very moved by this thought, and I picture myself on your pink and blue cards holding a bundle in my beak. But then I tell myself that I no longer want to be an image, a card-board copy, the everlasting symbol of power and consumerism in your species, the emblem of type-cast, computer-programmed children belonging to a family and class whose only belief is the status symbol.

I want to be the friend that storks used to be, pure and strong, as gentle and great as a blessing on your hearths and fields, your jobs and children's games.

But how long can our friendship continue if you change the rhythm of life by programming the number of children you have, thereby planning their disappear-ance? Don't you realise that, as we enter the twenty-first century, the future holds little more than an increase in the number of old people?

If I am to remain the symbol you have made me, what am I going to put in my bundle in ten or twenty years time? Believe me, this is not yet another anti-abortion lobby. Nor am I complaining about the possi-bility of being out of a job for the rest of my life. This bitterness comes from a friend who can see that the idea of growing and multiplying terrorises humans more than any other. Despite all fears for the world's future, children are still a sign of God's continued friendship with humans. Despite everything.

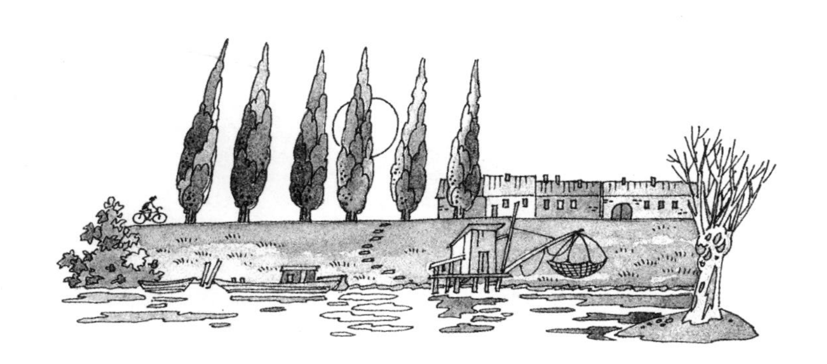

THE SPIDER

Dear Humans,

e spiders, as you know, are one of the most varied and numerous of species, venomous and non-venomous, large and small, each of us a perfect product of nature and divine craftsmanship. Since the beginning of time however, when you are faced with a spider, you treat us all in the same way. Whether we are venomous or not, you obsessively exterminate us all, your violence caused by fear and disgust.

A spider always gets you going, and often, when it comes to women, to the point of hysteria. You totally reject the idea of us spiders living under the same roof. And yet, despite your exasperating obsession with hygiene and disinfectants, you have to live with us whether you like it or not. Not even the most gleamingly polished, sterilised house is without its spider. It may be tiny to the point of being almost invisible, but it is there, ironically and happily tucked away in the smallest nook or cranny.

What makes you so disgusted? Why do you so indiscriminately exterminate us? Surely it's not because of your love of cleanliness. Every time you unexpectedly discover one of us, you have to admit that cleanliness and hygiene never have and never will be perfect in your exemplary house though you try to make it as antiseptic and airtight as a space ship. Just as well, because there are limits. The discovery of a spider-web is what gets you going. How can it be possible that one of us could have spun its web in such a short time?

Tell me something. Why is it that you, as a connoisseur, don't admire this humble web, the most fragile yet precious and gratuitous work of art produced in the animal world instead of destroying it? Or, before you destroy it, you could at least behave like the aesthete and hygienist you are and show a tiny bit of admiration.

Yes, even like a hygienist. Because a spider-web, be it outdoors or in your home and created from dust, is never "dirty". Tell me what, in your opinion, could be cleaner, more transparent or more precious than this work of pure beauty? Or what could be more necessary and useful for us? Does it not take talent, in fact genius to create such perfection? Don't you understand that you, the manic hygienist, are doing exactly the opposite of what we spiders do? What I'm trying to say is that all detergents, sprays and insecticides leave "dirtier" dirt on your carpets and tiles, chintzes and velvets, bedspreads and clothes, linen and leather, cutlery and china. I am talking about the amount of poisonous residue that you leave where, superficially, you mercilessly clean.

Contrary to your belief, we spiders do not like dirt and don't feed on rubbish. In fact we are hygienic, selective and even gourmets with refined taste. We are the creators, planners and builders of unique "cloths", each one specialised, each one as different as the spider who weaves it. The so-called "finished product", dear friend, originates from within ourselves, from the gift of creativity and beauty which Mother Nature gives us at birth as a dowry for life. In case of necessity, it can also be a defensive and offensive weapon. No, the spider-web is not just a delightful play-thing. It is also our guarantee of survival in a difficult, armed and deadly world shared by insects and other species of animals –

the same sort of situation that exists between various clans of humans, the most savage of all.

We spontaneously produce natural beauty on a small scale. Each of us is an active, self-sufficient component in this vast laboratory. The web originates inside our very bodies, our flesh and blood. There is only a minuscule amount of flesh and only a drop of blood, but it is enough and our ability to create architecture, sculpture and painting is far greater than any other animal.

I know what you're going to say: what about the silk worm?

I recognise this and we are, to a certain extent, friends but our work is both similar and different at the same time. He may be more rewarding but is also better rewarded than we are. However, look what has happened to his freedom... You have put him on the assembly line of intensive and profitable production. He eats mulberry leaves and turns them into silk, one of the most beautiful and valuable fabrics that humans – and only humans – use for their clothing. Once his production cycle is over, the worm can take it easy until the next season. He has long and restful holidays. At the end of his natural cycle he turns into a lofty and dazzling butterfly and lives a marvellous, albeit very brief "old age" visiting the finest "salons" in the animal kingdom with no fixed address of his own. I envy him for certain things. But when I think how immensely tiring all those changes must be, I return to the belief that my freedom is preferable. The spider would never be part of a Stakhanovite movement.

The worm's silk is spun into cloth, generating a staggering market by becoming a "must" for the well-to-do who buy these disproportionately expensive clothes. You don't use our "silk" for anything and therefore it costs you nothing. Even if it is useful for very little, we spiders pay the price. In many ways – the old "moral of the story" – this makes us more closely related to the cicada than to the ant. Our "silk" is only necessary to our own personal requirements: our nests, our food, our defence and to relations and communications between different groups. Because it is real beauty, it cannot and should not cost anything and cannot be used by any of you. It will never become a commodity for competitive trading and conflicts, a reason for open commercial warfare and terrorism.

As you take so kindly to the notion of give and take, I would like to offer you an "identikit" picture of us. It may not be of any use to you (although you do "weave" many a tangled "web"), but it will at least be a clear picture of how disproportionate the violence is with which you implacably destroy us the minute you see us.

There is a negative side to our spider-webs. As you know, commerce is built on the principle of "mors tua, vita mea". The fittest have always survived and we also live on as a result of someone else's death. We believe that some species of insect exist and reproduce for the sole purpose of nourishing us. We wouldn't dream of trying to fob ourselves off as pacifists full of universal brotherly love. We must also play on the dangerous and brutal swings and roundabouts of rivalry and survival for all the species. This is our destiny and we instinctively accept the law even if we don't understand it. If we could also occasionally have "feelings" as you call them, then I would say that ours would be one of great satisfaction every time a "giant" – an insect bigger than ourselves – falls into our net and can't disentangle himself to escape. He is captured and eaten alive. Our baby spiders are so happy when an especially tasty morsel comes their way.

Is that a disgusted grimace I see, my friend? You shouldn't do that. Don't be hypocritical as well as cruel.

You can't make allowances for extenuating circumstances like we can. Is there really any difference between our lots in life? Someone is always at the mercy of someone else. Does this not happen in your species as well as ours? Are you acting any differently from us when you go hunting or fishing? Don't you have "nets" like we do to catch birds and fish? Except yours are made of rock-hard nylon.

And when you triumphantly place your roast pheasant on the dining table (which perhaps you didn't even catch, but sneakily bought at the market), doesn't your wife praise you, don't your children clap you when they see their daddy as an heroic hunter? Haven't you just killed whatever came your way? Have you never been a poacher? The difference between us is that if you are caught red-handed, the most that will happen is that you are made to pay a fine and your game confiscated. While you use the heel of your shoe or a scrubbing brush to sentence us spiders to immediate death without ever telling us what crime we have committed.

I don't ever expect to cure you of your disgust for us. Nor do I think I will ever change your mind about hunting and fishing, when you reap all the benefits of someone else doing the killing for you by buying your game at any butcher shop. The great book of Law states that you are permitted to feed on all edible animals. With great sadness, all I can say is *Amen.*

I just want to ask you one thing. You boast of being a lover of beauty so you should be able to do it. Try to find the time to stop and look at our webs, stretched out in the morning sun or at dusk. Many people are starting to do this, so why don't you try too? If they are glistening with dew, those webs will entice and excite you, and perhaps you will be converted to friendship with us. If the sun should shine on the dew, our webs will be more splendid than the pearl necklace (makes no difference if they are natural or cultured) you gave your girlfriend in happier times.

My friend, all I offer you is this fleeting yet supreme beauty before you crush my masterpiece as you crush me when the dew has gone and the sun has set.

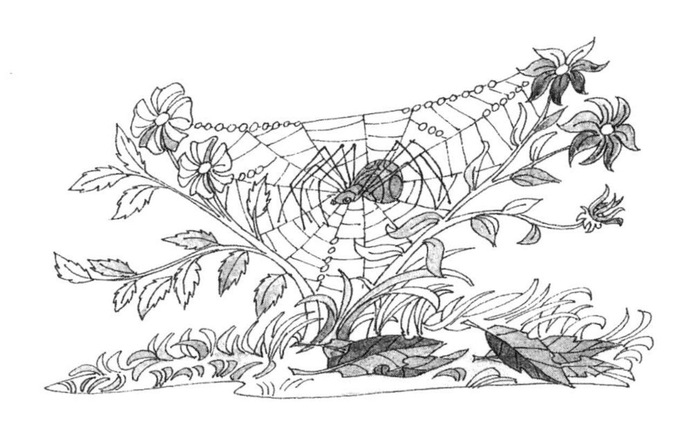

THE HYENA

Dear Humans,

on't be afraid. I am a hyena. I want to try to rid you of the prejudice you have always had against us. What I have to say is simple, but there is bitterness. You may find this hard to swallow, but I am tired of being your garbage collector. Wherever my species still survives we destroy the huge amount of rotting garbage that you generate. Your garbage, as well as that which occurs quite naturally.

I am not, nor have I ever been, cruel. In any case, what do the words "cruel" and "ferocious" mean when applied to an animal? Neither do I know how to smile or laugh as your saying goes. Is "hyaena ridens" just the name you have given to one group of our species, or are you trying to make me out to be sarcastic as well as ferocious? Besides, why would the blood-baths and catastrophes which you humans have caused throughout the world make me want to laugh?

"She is a hyena" is what you say about an unmerciful woman or an unforgiving man who runs rampant. In some languages where gender is used, the hyena is feminine as though trying to insult a male monster. Dear friend, why don't you update your knowledge of our species – all the other animals too, for that matter? I have never really understood whether you are our ancestor or our most odd and perhaps "mistaken" descendant.

Why is it that you insist on saying that I am "cruel"? Unless it is in self defence, when have you ever seen me, or when do you think you will see me, attack and eat a human or another live animal?

I live on dead bodies. This should be enough for you not to set yourself up as a judge of my behaviour. Don't you live off dead flesh yourself? Are you not carnivorous like we are? You can indulge your greed in the safety of the most hygienic and comfortable surroundings that have ever existed – in butcher shops. I live off the land, the desert, moors and forests, and only off the most revolting rotting garbage to be found. I don't feed off animals or humans (these days, almost impossible and unimaginable) until they have reached the end of their lives and their bodies are decomposing. I have a veritable feast after a battle (especially the old kind, with hand-to-hand combat or just a gun as a weapon), or after a fatal accident (somewhere uninhabited), or after a disaster where heaps of unidentifiable corpses lie scattered over the ground. However, unless I am dying of starvation, I never attack anyone.

As for the rubbish dumps with which I am always associated... I would like to ask you what, in your opinion, would you call some of your huge, miserable, swarming cities, not only in third world countries but also the suburbs where your so-called civilised people live? What would the vast slums in desperate countries be like without the compassionate help of the crow? Crows are more widespread in the East than we are, but I can tell you that, together, we are the only form of "municipal refuse collection" that always works and never asks for anything in return. Hyenas, crows, vultures... we all work for nothing. We are content with the "left-overs" quite literally discarded by humans and other beasts. Without these "dustmen", how could you avoid the plagues, contagious diseases and chain-reaction deaths throughout the world brought on by human greed and violence?

However, I am only an emergency "dustman". My survival does not depend on violent death resulting from animal or human struggle. I am usually content with anything that has died of natural causes. Vultures and crows, on the other hand, have a far broader range than I have. Don't forget, there is always some "left-over human". Why should I, the hyena, be merciful, sensitive and respectful when many of you no longer behave like this?

We live together and share this earth. But in what way do we share?

With your intellect and conscience, how do you feel about Mother Nature being so misunderstood, humiliated and insulted? She makes just laws even if we don't fully understand them at times. She produces her children, humans and animals, and when they die she consumes them, showing great mercy despite the fact they have have become repugnant. Quite simply, we hyenas collaborate with her to enhance her dignity so that she may take the credit for the hygiene and beauty. She has a clean record, the shame of death and destruction having been removed by us. We are never thanked. To give you an example, just think of the funeral rites performed by believers in the Parsee religion. As soon as they die, their nearest and dearest cut them into pieces, rub them with precious oils and place them on their ritual "towers of silence" as a funeral feast for the vultures and crows. In no time at all, all that remains are the bones.

I'm quite sure that watching us at work would make you feel sick. We know this, so we always try to work late at night so as not to offend you and to avoid ferocious battles with other hungry, desperate animals.

I am quite realistic but I need to get this off my chest. You don't have to come too close, but you at least owe me something resembling friendship and gratitude for what I do. Although you are "civilised", you face the same problems we do. As true as this may be, your hypocrisy will prevail. You wouldn't dream of placing a memorial plaque on our stomachs in recognition of the contents: perhaps the abandoned remains of an explorer or missionary. You are specialists in noble disguise and have often covered up the remains of your "dearly beloved" with great works of art. But even if you wanted to, you could never thank us. We will never get to know each other. Your disgust will continue to weigh on us from a distance. We will both be the last to die but we are separated by a cultural taboo.

And so you carry on, increasing the number of headstones on the graves and "whited sepulchres" in your landscaped burial grounds. Quite rightly, you also lovingly dedicate these sepulchres, be they "whited" or covered in flowers, to your most revered saints and benefactors of humanity.

But try to think of us occasionally. And also remember that a live animal is always better than a dead human or a false tomb. Although they are always full, there can never be any life in a tomb. They are all full, except for one. Luckily one is empty and it contains your faith and hope. You have paid the price for revering that empty tomb. You have been killing each other for centuries, thus providing us with food in the East or in Africa. And yet, I am very touched and even jealous when I think that one day you will be saved, you will rise out of your mortal remains only to return much purer and more alive than before. I am glad that this marvellous fate awaits you and that you at least have already won the final battle against death, an enemy for us both. Knowing that a man, the greatest possible man (whom you also call God) has won this battle because of his love for you makes my heart melt with affection for you.

But – and don't laugh now, don't become *homo ridens* – mine is not a totally unselfish love. I believe that eventually we will also have something proportionately similar to that which has been guaranteed to you.

I overheard a conversation between some old missionaries who had escaped a massacre many, many years ago in India. They were saying that death would not be the final outcome for us animals, with no exceptions, not even the gnat. One of them was also telling people that animals are the humans' "younger brothers" and "a smaller part of divine creation" and would, in the end, follow the humans into "Christ's mystery" (or Allah, or Jainism, or any sincere faith).

You know better than I do, than any animal does, who spoke those words and with what authority. You have to hope with all your heart that they are true. Because, basically, you like the world just the way it is, both here and now on earth and in the great, diverse and definitive universe where, fortunately, there will be no sorrow or trouble and no feelings of hatred for me. You like all of it the way it is and could not conceive of what it would be like, either now or in the after-life, without us animals. Perhaps death is just an ark like the one your Noah built, making its way through all the storms and horrors towards the bank of the great garden where we will at long last live as brothers.

I look forward to that meeting, my friend. No matter what, I will be there, waiting for you.

THE BEE

Dear Humans,

ou are taught in school that I, the bee, am of the aculeated hymenoptera species, living in a polymorphous community, and so on... In short, they will have given you the impression that I am a slave with no life of my own, locked into a cruel, spartan system.

However, I am sure that you have always appreciated me and considered me a friend, taking no notice of all the cut and dry text book definitions. You'd much rather admire my bright, glittering, golden, beauty. Although my little body is slight and fragile, its golden brown colour is perfectly radiant and I really do look more like a jewel than an insect.

And yet, dear friend, I am just an insect and nothing more. I am more of a robot (to use one of your current words) than a flying creature with reasoning powers. I don't mean that I'm lacking in what you call intelligence, or instinct at least. In all modesty I can say that I have such a high concentration of intelligence and instinct packed into the one centimetre of my body that I could teach a thing or two to you humans and other animals.

Don't get me wrong, I am not boasting. It's a simple fact. What's more, I also have that secret talent for grace and beauty which always catches your eye and melts your heart. In fact I can dance in the sunlight, performing a variety of perfect geometric figures as no prima ballerina could perform one of your classical ballets. I dance in the air and every dance is faultless. Contrary to how it may seem, that dance has a purpose. A purpose with no room for error. We dance to show the swarms of worker bees the whereabouts of flowering meadows, our nectar and pollen mines. We also dance to announce the miracle of falling in love, celebrated in the sunlight like no other creature of the land or air can do. Be it for love or work, we dance together in the air like exacting professionals, with both great fantasy and finely-tuned technical skill.

I know we are a strictly hierarchical society, relentlessly dedicated to our work and the hive. There's no shirking or any kind of relief. We are born hurriedly and work morning, noon and night. Most of us die within a few weeks, a month or two at most, burnt out and quickly removed to make room for a never ending chain of more workers.

Ours is a strict monarchic system. Only one queen reigns over us. Her all-consuming resistance survives for five or six years, whereas her tribe has regenerated by harvest time every year.

Of course there are males: the drones. Their sole purpose in life is to fertilize the queen and then disappear immediately.

Do you think our society is cruel? Perhaps. But what does the word "cruel" mean to us? If you want to understand our monarchy, our species, the animal world, don't ever forget that cruel is a human adjective. The standards which you may agree to apply to your life and feelings make no sense to us. Don't judge our life by your standards. We don't know anything about strikes, we can't imagine any other system for living together except that of the hierarchy and monarchy which you often scorn and reject.

This is a natural system with a specific aim – a system which obeys nature. There's no point in wondering, as you usually do, whether it is worthwhile living like this when life only lasts a few weeks. Despite the hint of hypocrisy, don't you admire heroes who sacrifice their lives for a worthy cause and die at a young age when compared to your average life span? Why do you feel sorry for us if we give all we've got for the sake of the factory? In proportion, our workers weigh one gram compared to yours whose body weight is forty, seventy kilos.

When the queen is old or has a stronger adversary, she does not conspire to keep her honeycomb palace. She leaves peacefully to set up a new community elsewhere. She doesn't want to be pensioned off, retired like all great managers, and never loses her temper or makes others angry.

When fertilization has taken place and the group's life is guaranteed, the males are abandoned to starve to death. This is the plan and, unfortunately, it is carried out. Nothing more to be said. This is another law which confirms and guarantees the continuity of our realm. Everything always works perfectly.

Survivors can only rest during hibernation. No sooner have they fallen deeply into this sweet slumber than they are ready to wake up again with the coming of spring. But in the meantime, between sleep and dreams, we have as much time as we need to think – unlike yourselves – about the moving stories told about our species. As soon as the sun has opened the first acacia buds, we are ready, if we are still alive, to start all over again. Between sleep and dreams we talk of what our future holds, looking forward to the rays of the morning sun. We produce everything we need and are self-sufficient. Our rules do not permit us to ask for outside help. Then suddenly and usually carefully and lovingly – you, humans, intervene to be sure that your supply of honey reaches the market, temptingly packaged but almost always reduced, pasteurised and robbed of its precious vitality. We don't steal anything from anyone. We simply gather the Creator's gift to the flowers which he offers to one and all, the good and the bad. Because wild flowers are hardly ever picked, our food is not stolen. On the contrary, we are "rewarded for our labour" as your brother Christ said. We earn the gift offered by the flowers because we are the ones who transform it into the "nectar of the gods" as your ancient poets have called it. Who but us could take honey from a flower?

We only give, asking nothing in return from you humans. Isn't this enough for you?

No, you are not satisfied. You are never satisfied. You steal and adulterate our honey. Out of sheer ignorance, you often destroy our hives, shock us and frighten us away to a make-shift home. The end result is that you adulterate this earthly and solar gift which only we are capable of giving you.

At this rate, we are also on our way to extinction. As we look at the toxic clouds of your industrial waste, and feel the change in the flowers blooming on plants ruined by thousands of poisons added to fertilizers which no longer aid but have become enemies of the soil, water and the air, we feel that we want to teach you a lesson. In fact, we are doing just that. Our sting is like that of the wasp. Stings on some parts of you body can be fatal. Some have already died. Your rough approach to our hives, rudely interrupting our perfectly timed work, will upset us and change our tactics from defensive to offensive: a small, fragile and frail army of creatures, no bigger than a drop of water, peace-loving by nature and vocation but armed, to use your words, "for self defence".

Humans, you can't go on forever being deceived by our appearance. We will always try to do our best for you, time and again. But you can't just "touch wood" and hope that everything will be all right. If you carry on with your selfish behaviour, polluting the sky, the soil, water and the flowers with noxious fumes, you will very soon sign our death warrant. This will make things much worse for yourselves as well. It will mean that the earth is no longer inhabitable and the air no longer breathable. It would certainly be bad luck for the overweight gluttons if they were deprived of their greedy privilege and had to do without their honey at breakfast. Without honey, how could your most refined and choice cosmetics guarantee silky skin and shimmering bodies?

Take notice of what I'm saying, my friend. Let us at least teach you to be strict and scrupulous about your work. When you are trying to show your love for others, learn the drone's ability to give the fruits of his labour freely. Learn from us that, first and foremost, the sweetness of friendship nourishes the heart and soul more than the flower's sweet honey can ever nourish the body.

We are happy to set this one example.

So that we can survive and live with you as creatures who have rediscovered their original happiness and innocence.

THE ANT

Dear Humans,

here's so much slander, don't you think? There are so many lies told about us! Not to mention the praise! And yet, outside my tribe and the ants' nest, I, the ant, have no importance. Work *unquestioningly*: this is our motto, put into effect every day of our lives without argument. We ants work until we drop. Mother Nature's plan permits nothing else. Depending on the ups and downs of the seasons and harvests or the mood of our top-level management, the only advantage, if you can call it an advantage, is that we can more or less count on being able to hibernate.

You know better than I do about the lies: there is no escape from being habitually compared and contrasted with the cicada, making her out to be my arch rival and enemy. Despite appearances, this goes against me as well as her. It serves only as justification for your industrious initiatives which often lead to your cynicism and greed increasing at the same rate as your accumulation of wealth.

First and foremost, however, I want to ask you not to credit me with ideas, opinions, limitations and responsibilities which we animals do not have. Writers of fables and moralists, who, for better or worse, are the same, including the one writing about me right now, can't help turning anything an animal is destined or programmed to do by nature into a vice or a virtue, guilt or innocence.

Having said this, I hope you won't be offended if I compare you with an ant. That is to say, according to your codes of conduct, into one who suffers from the syndrome of intensive gathering and stock-piling. In short, a miser, regardless of whether you are rich or poor. I must remind you once again that we are neither misers nor spendthrifts no matter what your lovely fables may say. We are quite simply predestined and programmed to do what we have always done since our species began, and do it as best we can.

Don't type-cast us in a role which you have created, directing us to play our parts mechanically without understanding the secret and mysterious flicker of light which, if you don't mind, could be called our "soul". Quite frankly, you foolish and greedy squanderers, all that we ask of you is not to make excuses to justify your actions. If we agree on this, then it will be possible to understand one another and live happily together in this world.

Even if you don't take any notice, we ants still believe that we set an example. You could even say that we are your sociological and moral role model. As small as we are, we are to a certain extent also a definite entity. I might as well whisper this, in fact (and it's not absurd) I will tell you in silence as I don't think you can hear my voice anyway. Judging from past experience, I believe that you have not yet reached the stage where you can take in what we are saying and want to make you understand. Because our sister, the cicada, can sing you would think that she would be able to tell you everything with her constant chirping. In truth, not even she knows whether she is somehow communicating with you.

We are content with our lot. We believe in obedience, discipline and work. I know that we are not outstanding enough to stimulate your imagination. How can you attach importance to someone whose voice is not audible to the human ear, to someone who you often can't even see because they are so small? We are the "nobodies" of the animal kingdom, the supplies department. Who could be more anonymous, more of a "nobody" than the ant? One ant, or one hundred, one million ants... who sees them if they are not looking for them? Without a microscope, who could tell the difference between a male and a female, the chief ant or the worker?

And yet, ours is a great republic, ideal, perfect, constant and safe. None of us ever dies of hunger. There are guards, explorers, gatherers, carriers, store-keepers and distributors in our republic. You who are so thrifty, hard-working and, in your own way virtuous, often describe an active person as someone who "scurries about like an ant".

It may be true, but it is praise given to us out of pity. I wonder if you have ever realised the price we pay to guarantee a home, food and freedom for our children, especially in winter when there's not much lying around and life grinds to a halt during hibernation. Hibernation has also been designed so that we don't die. When and if they notice, only your most well-mannered children show their amazement when they watch us with innocent, admiring eyes. Then you are made to understand that our efforts can, on a smaller scale, be compared to Herculean tasks. They are always incredulous at the sight of a tiny ant carrying a crumb often two, three, four times bigger than she is, never stopping until she has placed it in its proper position in the secret, underground silos. Your children always admire us as models of a conscientious, responsible labourer, working in solidarity with its entire clan.

But I am not here to be praised.

We ants would be content if you stopped casually

and cruelly exterminating us when we leave the nest to get some sun or approach the crumbs that have fallen off your table. You almost always sentence us to death. There is no reprieve. You hear your children say: "Dad, I've seen ants in the house. Where's the insecticide?"

The only ones to praise us have been your fable writers. But you are always right. You can use your reasoning power to juggle with heroics and crime, rights and duties, work and strikes, loyalty and betrayal. None of these things exist for us and we don't understand them in the slightest.

Many rich or greedy humans take offence when they are compared to ants. In public, that is. In private they are pleased. This is because you are more miserly than far-sighted, slaves to your treasure rather than its masters.

As yet you don't say that we are as mad as the cicada, but it won't be long before you do. Because the relative development of your economic well-being, especially in the less poor countries, creates the tendency to earn as much as possible to fulfil your every wish. The extent of your greed leads you to be both extravagant and wasteful. But heaven help anyone who should say this! You consider us to be frenzied slaves, driven by ultra-sound rather than reasoning brains. It won't take long before your children consider us to be anything but a role model.

When you visit the pyramids of Egypt or see them in advertising brochures, you think of the slaves who lost their lives building those stupendously imposing works and whose names have never been recorded in the history books. What you don't take into account is that in conclusion, comparatively speaking, when you have reached a certain level of power and authority, you still have your slaves who do the same thing day in day out. The only difference is that these slaves have a name, a union card and a police record and appear in the tax register.

Although we are programmed, electrically charged and therefore mini-robots in every way, what would you do if you were to discover that a flicker of identity, of "individuality", was gradually growing in us?

Before you answer that, dear friend (you might be part of a respected ecological movement), just stop. Before you squash or poison us, stop a minute. It could be that we are or could later be evidence of the "evolution of the species" as you call it.

You may not believe it, but we are also important, even more important than the pyramids of the Egyptian pharaohs. We are as important as dams, motorways, sky-scrapers and huge cities. Our nest is our living space, our defence, our home. So don't get rid of it, destroying everything with one contemptuous kick. After all, we are the cleaners of the impenetrable corners in your home, your world. You, on the other hand, unrestrainedly and unscrupulously dirty your home and ours.

We certainly don't expect you to live in a "system" like ours. You must be careful not to end up in the oppressive and shockingly overcrowded conditions already existing in many of your greatest and most admired cities. We are extremely comfortable in our ants' nests. As you have seen, we are also at home in your "nests" where we can raid the rubbish that more often than not you don't even see. You are the ones who no longer feel at home. You feel suffocated, existing like complete strangers forced to live together. Don't sentence yourselves to live by our standards which are the opposite of yours despite their seeming similarity in your eyes.

Each of you is an individual. Thrown together as you are, be careful not to become just a crowd. We would like you to be a people. A people who are the friends of nature, creation, the animals. Even the ants.

THE CICADA

Dear Humans,

am song, pure, carefree song. But whenever I see you approaching the branch where I am giving my very own private summer concert recital, I am so startled I almost sing off key. Although I know that, deep down, you love beauty and sunshine, summer and song, time has shown that wherever you go you almost always bring the risk of my demise with you. Who but you have taught your children to "have fun" by destroying ants' nests without giving it a second thought? Once I even saw you enjoying the massacre as you watched it through a magnifying glass. Who devised the cleverest gimmicks and traps for your children to make surprise attacks on cicadas, catching us off-guard, drunk with song in the summer heat we love so much? How can we defend ourselves from the grasping little hands of your young when everything, even our tiny bodies, our reason and prudence, is lost in the zithering sound?

I have also always figured in your children's reading books. But I came in through the servants' entrance as the "idiot". Unlike the ant who, in these fables, wears the crown of "wisdom", foresight and industry, I am always the absent-minded, irresponsible good-for-nothing. Except for my song.

Except for my song! Isn't my song enough in itself? Or do you think it's just a gift, a good enough reason for my brief life, insignificant after a few repeat performances of my gratuitous and foolish concert?

You don't even need this much to justify your pedantic fables, your useful morals camouflaging the meaning of foresight and justice for you and your family. They turn me into the symbol of foolishness, a mad creature intoxicated by the sound of its own trills, who chirps with no thought for the consequences of this distraction nor for the few enthusiastic, sincere listeners. Even when a poet – the only human prepared to understand me – admires the sacrifice of my life for song, there's still no change in the reading books, the fables and tales told specifically about me by humans. "Ant" is often a word used as a form of unconditional praise. "Cicada" is an insult and rarely, as I said before, a sincere commendation. You are all mixed up. Prohibited subjects like personal possessions are confused with the ant's merits like saving and work and the cicada's foolishness in not putting anything away for a rainy day, "loafing about" (this is the exact phrase; I have been assured that this is what is written in many children's books) all day long.

All I really want to know is have you ever realised just what my lyrical season costs, or understood the natural mystery of my charitable performance and my short life span? What do you know about my thoughts (or don't you think that I could have anything equivalent to yours) and the ideas that cross my mind and antennae? This is how I am made. I couldn't be any different from the way I am, and it is unjust to say that I have a choice as though I had the alternative to change my way of life.

I have stopped objecting to the same old story about me being the good-for-nothing and the ant being the far-sighted one. I often have friendly chats with my ant friends during rest periods from our respective work, and I can assure you that they are also tired of it. Dear humans, don't present us as though nature and the Creator had made us enemies. They have made us differently, that's all, each of us with our own purpose and appearance within creation's wide range of animal life. If a famous artist lives for his art to such an extent that he leaves nothing to his children – and he's wrong, very wrong, because, as a human, he has a greater duty than we do – why do you still build monuments in his honour? No-one has ever dedicated a monument to me. Not that I would expect one or even like it, but this does not release you from the obligation to correct your pedantic fables once and for all. It is not as though the value of art is not equal to the merits of foresight, even among animals.

We are finished when our song has finished. Our gift is total, absolute and, in the end, silent. No-one sees us die, and no-one knows what happens to the zithering little cicada when she dies. Then, when summer comes around again, more cicadas appear to sing their brief, unique and gratuitous song to you and all creation. Where they come from or where they are born is of no importance. How can you go on saying that our life is meaningless and our song is futile? You who are slaves to the superfluous, obsessed with ownership, why can't you understand the beauty of our sacrifice? Or is it that you don't want to, because for a fleeting moment, when you do think, you understand that our ephemeral life leaves behind an echo and a reflection of the happiness that existed in the "garden of delights" when the world began.

With a total absence of forethought, I would like to ask you in all sincerity if you and your children really think it is of some use to dismiss me in favour of the ant? Are you sure that you are offering your children a complete education, a genuine and beneficial moral standard? How can this be done if you make one creature out to be a "devil", only to make a "saint" out of another? Do I or don't I have a place, like the ant, in this marvellous world that is so hospitable to everyone?

Do you really know what goes through your children's minds when they hear the same old fable about the cicada and the ant? Are you at all able to understand that many of them, even if they wouldn't dare say so, are on my side and admire me more than the ant?

Like the ants, the lizards and baby birds, we cicadas fully understand the instincts of these children. They watch you and imitate your example. Judging from experience we can tell you that, even without realising it, they are just as cruel to us (as well as to each other) as you are.

Most of all they want to have fun with us. Their smooth little hands grab us violently. Elated by this capture, they crush our wing-cases and diaphragms, wings and eyes, squeezing us until we suffocate. As soon as we and our song are dead, they crush us with all the contempt and anger of one who blames a toy for being broken. "See if you can chirp now," some of them say as they look at what they have killed, staring at it with pitiless eyes. It doesn't happen quite so much these days, I know. But it does still happen. Are these little angels like you or worse than you? I think and hope they will forget your lesson and find another way in their hearts, not the path trodden by torturers hiding behind the shield of hygiene.

My human friend, don't be offended by the bitterness of my words. Can't we at least be happy that more and more of our friends enjoy mutual love and respect? I know many, and I often joyfully relive a beautiful story of friendship which I would like to tell you. Eight hundred years ago, there was one unforgettable lover of cicadas. As poor as he was, he was full of song, friendship and love for all creatures. It was no accident that one of your great poets called him "the sun". "Sun": the fiery source of our song and life.

One mid-summer day, this extraordinary poet challenged one of us to a singing contest in praise of the Creator. He chose his tiny adversary deliberately because he had listened to her sing for days on end with increasing admiration. As he was also a great fun-lover, he wanted to enjoy that strange duel. It was a memorable contest and apparently silenced all other voices for many hours near the little church of St Mary and the Angels in the Assisi plain. For a long time he was what you would now call an excellent singer/songwriter, a minstrel. His mother was from Provence, the French region famous for its troubadours, and her blood ran through his veins.

Our little sister cicada accepted.

And won.

This amazing episode has been recorded in books telling the story of Francis. However, after a week of contests repeated for many hours every day, I believe that the only reason my courageous ancestor was happy about her victory was because she could sense that Francis, who was considered "small and stupid", had only challenged her to honour her, to let her win and, what's more, that he was truly convinced that he deserved to lose.

No-one knows what those two said to each other during the intervals at that great concert. Perhaps nothing at all. What need was there for words? Aren't those who love to sing quite satisfied with the sound of their own song? Not even the cicada's name is of any importance. Except for song, her adversary's great love made him poor and the fact that she did not even have to give her name must have made her happy.

If you really are my friend, call me "sister cicada" as a good brother would, in honour of that challenge.

THE MOSQUITO

Dear Humans,

ast August bank holiday my sisters and I miraculously escaped being massacred by some tipsy people merrily drinking all night long on a terrace overlooking the sea. They were singing, drinking, dancing and eating. All that noise had irritated us and we probably went too far with our so-called thirst for human blood.

And yet, none of those people lifted a finger to shoo us off or squash us as normally happens. Instead, those happy holiday-makers exterminated us with infallible torture instruments, without dirtying their hands with our blood.

You have perfected the fastest and surest means to do away with us: deadly, silent lights and smoking coils which quietly take immediate effect. Unfortunately, many of us swarm towards that bluish light and deceptive heat like moths. It is grotesque but true. We are victims of light, the one thing that has always attracted and fascinated us most of all (together with the smell of your skin and blood).

In my own case that evening, it could be said that I was perhaps glad of the opportunity to join in your party. I always think that people at parties are happy and therefore good, or at least kind, never whimsical enough to kill even by accident. As soon as the party begins, we dash towards that smell straight away. And straight away we meet our death. But what is death for a mosquito, or whole swarms of mosquitos, in all the hubbub and humidity of those hot and sultry summer nights? No-one could possibly notice the tragedy of our sudden death. Once we've entered the bright whirlwind like living confetti, there's no escape. I suppose it could be said that the horrible smells you have invented to keep us at bay, not to save our lives, are kinder than those infallible machines, our funeral pyres. When we are eliminated by a swipe of the hand, a handkerchief or a fan, all that generally remains of us is a spot of blood and a microscopic stain on the hand or the wall.

All that can be said, dear friend, is this: you have made incredible progress in hygienic extermination, reaping the benefits immediately. A whole field of chemistry is now dedicated full time to insecticides and pesticides. You never for one moment consider that your health and hygiene causes our savage elimination. You are quite explicit when advertising this field. Using the typical stupidity and absurd nonsense of advertisements, many companies daringly present a dilemma. The company's name is followed by the assurance of death. "PGM... or death!" The exclamation mark says it all with no need for further comment.

You may say that we do annoy you and this is a real problem, not a passing whim. You are right, but much depends on the fact that the natural relationship between you and us has been altered. We have become poisonous to each other. The least cynical among you admit that the "ecological balance" must, as always, be defended and helped by political undertakings, environmentalists and enthusiastically dedicated groups of young people. But what good can these sincere beings and liberal groups really do with their outbursts? Will there ever be a human who comes up with arguments, reasons, enough credible evidence to launch a campaign to save, for example,... the mosquito? No, I don't accept this hypocritical contradiction either. I know full well that the mosquito has always been and still is the carrier of epidemics, infections and catastrophes. Together with certain species of fly, we have become the unmistakable carriers of many diseases. It is a constant reminder that the recognised difference between our overall usefulness and our actual harmfulness has become an irreparable fracture. Of course, all animals have a purpose, or a vocation as you would say. However, in humans as with animals, power, cultural differences and sometimes even religion prevent us from relying on this role and balance so we can live together and improve the quality of life.

My friend, don't forget that the "human animal" is the proudest and most treacherous. Regardless of the consequences, he respects only those greater than himself, those who have more power to guarantee well-being.

If I wasn't so small, almost blind and with badly singed wings, I would like to see how you'd manage in a test like mine. The human race always considers a test to be cruelly significant. If we treated you the way you treat us, it would mean genocide, the "final solution" aimed at your "society" alone. And if, like me, you escaped planned extinction, I think you would find it difficult to resist the temptation to treat others as they intended to treat you.

I am not preaching you a sermon. I want to tell you something very simple: treat all creatures with respect and without prejudice. Defending yourselves from the real danger we represent to you is quite legitimate. But you know better than I do that there's a big difference between self defence and unnecessary extermination, a vengeance to the bitter end against defenceless inferiors like us. Vengeance is always blind and makes the victim as murderously hateful as the perpetrator.

From a health point of view, as things stand today in all the undoubtedly civilised countries with all the hygienic certainty at your disposal (apart from pollution and nuclear threats), I believe that you could consider the mosquito as an "annoying presence" and not an "enemy".

Let's be friends again, abounding in a friendship that signifies "disarmament" for all living beings. And to sincerely "disarm" the mind and the heart, you need to see clearly, without your vision obscured by fear. One of your scholars wrote: "at three hundred metres an enemy is a target, at three metres he is a human being". The distance is more morally symbolic than real in physical terms, but if it is not reduced we run the risk of improving our strategy of attack rather than defence. You will be violating your outstanding science and we will be regardlessly developing our instinct to take our daily supply of your blood which, unfortunately and for reasons I don't understand, we continue to enjoy.

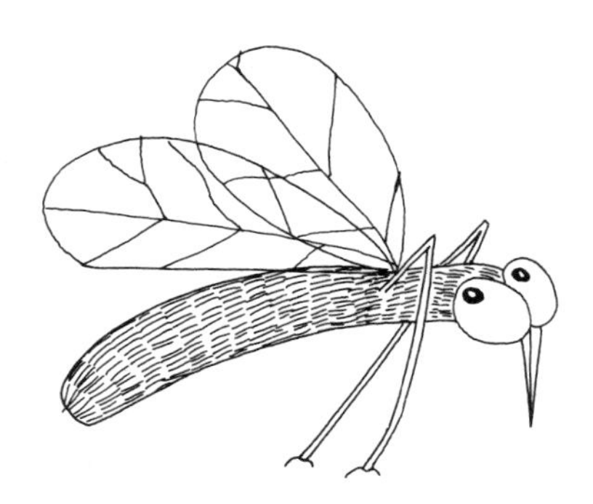

THE DOVE

Dear Humans,

or the first time in history you should, I think, begin to pity us doves. We get a mention in practically all your religious and secular literature. You have made both male and female doves into the symbol of peace, of everything that is pure and glad tidings of all kinds. The first time the Holy Spirit was clearly and solemnly manifested to you Christians, it was in the "form of a dove" hovering above Jesus Christ as he was being baptised by John in the River Jordan. These signs and metaphors do me great honour, but I hope I don't shock you when I say that I think you are meddling in matters of humanity and divinity, with real and symbolic things. I hope you don't mind me saying that you will be forgiven for much of this because you don't know any better. For the time being, let's not concern ourselves with the holy Book. As unpleasant as it may be, look with greater intent at current events instead of great historical events and try to learn a lesson from them.

Begin by remembering what the scientists and naturalists have to say about us doves: we are the most lustful of all the animals. Nonetheless, we are still stuck with being the symbols of purity, innocence and crystal-clear morality. Symbols and messengers of peace no less.

Far be it from me to deny or argue with the importance and meaning of all these things connected with my image. I am well aware that "image" means a great deal in this transient civilisation, and I don't deny that it sometimes has a positive effect. But, by the same token, we doves would not want to be angrily reproached by Christ because of this hypocrisy.

It is no doubt an honour that Christ had no trouble accepting the symbols, similes and proverbs presenting us as the image of purity and sincerity. But how could he possibly have rejected them when the biblical people knew of the important role I played in the first "glad tidings" in biblical history, when I announced the end of the universal flood? And how could he dismiss the image itself when, in the Jordan, it showed him to be the Son of God dedicated to the salvation of all living beings – humans and, dare I say, animals? He had far better things to do than waste his precious time on abstract thoughts about the choice of symbols for his plan to free humanity. So when he wanted to give his followers an example of how to behave towards one another, he said that they should be "cautious as snakes and gentle as doves". What greater merit could we have had in your religious books?

But this is not the reason for my letter. What I really want to do is help you understand the far simpler, more immediate, dramatic and, in fact, tragic reality. I am inviting you to listen to the truth but, as you are about to enter the third Christian millennium, I don't know whether you will be able to accept it. Luckily, I see that this truth already fascinates many of you and is very close to your hearts. Firstly I have to say that I am not actually a dove, but a carrier pigeon: a species which is becoming more and more rare, doing a job that is becoming increasingly more difficult with the continuous risks of muddled changes in course. I even risk my life on many occasions. I decided I wanted to do this job when I was very young. With the development of social, technical, industrial and cultural communication all over the world I thought it offered excellent career opportunities, I have to confess.

As I grew up and learned more about this longed-for career, I began to become very frightened by my choice. Your terrifying computers and the proliferation of these technological monsters which now permit you to live, see, know, buy and sell, cure and kill at the stroke of a key have already destroyed me even if I continue to live and travel.

At first I was happy (I was still young) to play and share the air space with all those impulses, signals and codes. I considered them to be playmates, darting between them unharmed as though I was being shot at with corks from your children's toy guns. Then I realised that computers are in no way innocuous. Their ability to deceive or save, build or destroy depends on who is using them and for what reason. My proverbial innocence was finished the day I discovered this.

My last job was in a nuclear armament centre. So, besides pleasing my mother who always worries about me, I now count my blessings for having found my present position, and very limited radius, with one of the last utopians: a breeder of carrier pigeons. When I am not doing the postal rounds, I instruct barely a dozen other trainee airmail pigeons. Although they are decreasing in number, we supply many groups of people who are fed up and infuriated by the deadly interference which riddles space at both high and low altitudes. They prefer to communicate via us (and have thrown away their gossipy walkie-talkies).

The work isn't hard, there's plenty of bird-seed and lots of free time. It would be an ideal existence for a retired pigeon. But not for me, in the prime of my life. However, as I wander around, thinking – and eavesdropping – I have come to the realisation that you have no more secrets, privacy or time to yourselves. Whether you like it or not, you are always receiving information, news both good and bad enough to give you a heart attack. Therefore my concern for you is sincere when I think of what God said through one of his prophets in the bible: "The world is desolate because no-one reflects in his heart".

"Reflect in your heart" are the most beautiful (you would say "inspirational") words of advice ever written by the very God who appeared above Christ in the form of a dove. The time for reflection has indeed come before it's too late for us and for you.

My grandfather twists his yellowed beak and says that your information technology and methods of tele-communication are as infinite as those of Divine Providence used to be. But he also adds that, at this rate, common sense and religion will cease to exist. What pleasure can you get out of not having anything left to search for, to guess at, to look forward to and challenge? Although a double-edged human quality, curiosity is primary and irreplaceable. What would you have left without it? Wonder, amazement, true unawareness of the truth, of beauty and scientific discovery searching for the necessary means to save humanity rather than wipe it out forever: where and how will these most precious possessions end up? At this rate, however, there is no danger of you becoming either inwardly or outwardly idle, no longer curious, no longer moved, no longer resentful for no reason. Everything will be immediately and simultaneously contemporaneous to each and every one of you. Your habitat will be an evil "Sea of Tranquillity", soon to be as sterile as the desert sea of the same name on the moon. Dear humans, you may think you know everything there is to know as you stand there yawning under the increasingly more fruitless tree of good and evil. You think that you are already on the point of being "like God", a god yourself. But you forget who it is who made this promise to Adam and Eve, and what happened after. How can a repeat of this event be avoided?

You could do anything you wanted, but in what way and for whom? You could even take steps to avoid wars and live peacefully. But perhaps you can only do this if you fool yourselves into thinking that you are living in the "garden of delights" like the first people on earth. Instead it is the desert inside you – it is already spreading – that will eventually surround you. That "garden" is nothing more than an endless forest of television aerials and lethal smokestacks.

As you know, I personally was not in Noah's Ark. I am well aware of the facts, but that was certainly our first and greatest moment of glory followed by that of the descent of the Holy Spirit on the kneeling figure of Christ in the waters of the Jordan. Brother human, it is for this very reason that I hope to perform a repeat of that first peace mission between heaven and earth. Before I get too old and lose hope, can you show me a sign of recognition? It would mean such a lot to me. Because the truth is that since the first olive branch was carried by the dove to all peace-loving people on earth, there have been very few recipients of our messages whose reactions have given any hope for lasting peace and justice.

These days I only fly for my own pleasure, jogging, as you would say, at mid-altitude with plenty of time to think. Sadly, I have realised that I am eternally indebted to you for providing me with this opportunity. After all, planes fly at a much higher altitude and fighter jets don't shoot at pigeons...

Our future, then, should be quite rosy. But it isn't, not at all. Because many of you don't "practise what you preach": ecologists at home but killers at work.

These people would hardly call themselves enemies of the animals, with their canaries and budgies in the living room or a goldfish in a bowl. But when it comes to city pigeons – this is the main point, the scandal, the Nazi-like terrorism – they concoct and carry out infallible schemes for extermination to the bitter end. They say they have good reasons. Pigeons are filthy. They pollute and ruin fine monuments and soil cities of great beauty. What's more, they reproduce and multiply before your very eyes. How dare they? But you are ready and waiting with your two great "final solutions to the pigeon problem": cyanide or contraceptive pills.

Don't think that I am not aware of the serious nature of the problem. It does exist: we are a threat to the fine appearance of monuments and hygiene in the cities. Something has to be done. That's the truth of the matter, and it's very hard to swallow. Let's not even mention the souvenir photos of your children performing the quaint, reconciliatory custom of feeding the pigeons. There are still times when you consider the need to teach your children, in the name of pollution, to throw us poisoned seeds (to once and for all solve this threatening problem).

Why don't you hygienists, ecologists, psychologists and pacifists find the means to teach a species such as ours, for example, to live in the country instead of the city? Would it be a scourge on the farmers, or would there be more variety in our feed with less damage to the crops than to your urban hygiene? Yet again, we share the same problem. As you have reached the point where you'll soon be able to spend your weekends on the moon, you should be able to treat us reasonably with practical plans that don't necessarily involve our regular extermination.

I'm not telling you all of this so much for myself as for my countless brothers who are destined to die.

Every time I agree to work for a friend, either the sender or recipient, I realise the privileges of being a carrier pigeon. Some jobs offer the opportunity to enjoy the sky and gaze lovingly at the earth. The food is much better at that height: flying insects, not to mention the great variety of seeds on the plants below. And as I said before, there's the prestige. Despite everything, I still carry written, signed, personal messages even if, more often than not, in code. This century has seen the end of one of humanity's greatest treasures, the exchange of love-letters. The beauty and passion of these expressions of love have been defeated by the telephone, so, for now at least, I am timeless and still the most romantic of postmen. Just think of all the bereavement which could have been avoided if Abelard and Heloise or Romeo and Juliet had had a friendly pigeon like myself, instead of risking the tragic consequences of meeting face to face. I could have been of great use to Shakespeare.

THE DOLPHIN

Dear Humans,

lthough it worries me somewhat, I am beginning to believe that I, the dolphin, will soon achieve a goal: I will probably be able to communicate with you. If not with the same words, then certainly with the same code from which, in time, we will develop a real common language.

My father has always told me that if humans understood our sounds as we, for some time now, have understood what they are saying then we would have already easily reached the point of communicating and talking. However, my father always warns me about being in too much of a hurry and to think about it carefully, so neither you nor I are under any illusions.

I like being with you, performing for you, entertaining your children. In return for my aquatic acrobatics, I enjoy the good quality fish and other strange but nonetheless delicious food which only you know how to make.

Not long ago, just for fun I frightened my human trainer by telling him that if he keeps up the present work pace with so many people crowding the stands that they don't know where to put them, I will no longer be content with payment in food. Like a human I will ask for money, administered on my behalf and for my children by an honest human who will take care of our expenses if, for some reason, I am not able to do so myself. I had always thought that my trainer was genuinely fond of me for my own sake but, you know, he was so taken aback by my request he almost had a fit!

So I was the only one to enjoy the joke. It didn't take long to calm him down and everything went back to normal. However, since then I have continuously questioned the matter of your "domesticating" us animals. Although you have never openly discussed this with me, you know better than I do that it's not a problem to be taken lightly.

The thought first crossed my mind when a circus was set up for a month or so in the public park near my aquarium. By repeatedly making vertical jumps out of the water, I was able to watch at least half the sideshows. They certainly were marvellous... but very distressing!

Seeing those living mountains of flesh and blood, the elephants, standing up and balancing on their great feet, whipped as they pitifully danced to receive a sweetie filled me with indescribable sadness. Then I thought about what I do every day in my aquarium home in front of thousands of spectators and I decided that the only difference from the elephants is that I am lighter and more agile than they are. But this is no consolation. And now and then I regretfully think about the freedom enjoyed by real dolphins in the distant, open waters, greedily enjoying the left-overs thrown overboard by the passing ships they greet with their nimble dancing and splashing. Whether performing alone or in pairs, their life is a fanciful and rich tapestry of companionship and happiness.

It is true to say that I am not exactly a slave. The aquarium has a border, but there are no limits below the surface. I could leave as and when I please. But I have never seriously thought about it, nor would I ever do it because of the great excitement I provide for so many children with my inventive tricks.

Perhaps it is mainly because of them that my family and I have stayed here. By doing this I realise that I am educating (or miseducating?) the children and giving them a good idea of the friendship and proof of the solidarity that can exist between humans and animals. However I am also showing them that, as always, animals are dependent on humans.

So, my friend, shall we once and for all put the meaning, for both of us, of the verb "to domesticate" to the test? For starters, don't you think that the word itself is very inappropriate and a trap? Who is the domestic one? Who lives in a house? You have one, two, perhaps three beautiful, comfortable houses to live in and to reflect your wealth. My house – I have never forgotten this – is the sea, not the aquarium, not huge olympic pools. In fact, I was born in freedom in the ocean, and it is there that I would like to die in freedom. Dolphins may well enjoy leading or following a ship that provides an excellent snack. They are obliged to show their gratitude in their own way by the merriment of their aquatic dance. But it is not the same thing as being in permanent "service", obeying every whim and order given by owners and trainers.

However, compared to the elephants, I consider myself lucky.

More so when compared to lions. It's true to say that human rather than animal standards are not applied to lions. They are not dressed up and made to look ridiculous with gaudy belts and ribbons, feathers on their heads, howdahs on their backs and other flashy ornaments. But, whatever dignity the "king of the jungle" may be permitted to have, he is tamed, confined and trained to pretend he is docile even when he is ferocious, or to pretend to be ferocious when all he really wants is a peaceful snooze. I'm sure you've seen the little man (usually dressed in red with funny gold trimming, holding a ridiculous little whip) sticking his head inside the lion's gaping jaws. Well, this is when I pray to the Creator of animals to show that he does not fully agree and will teach a lesson (the choice of lesson is his) to the silly little man, the lion and the children in the audience so they learn that they cannot continue with a "show" which should have been wiped off the entertainment map long ago. And when it comes to bullfighting... these days its followers are usually only irresponsible fanatics who drug the poor bulls, the massacre paid for with the revenue from overpriced tickets.

You probably think that I should mind my own business. But I am saying all this out of conscience and pity in the hope that you will understand. A show should never consist of animals – or humans – with a tight reign kept on their freedom. Even if initially out of friendship, they should not be forced to earn a crust by performing doubly difficult tasks which they instinctively or naturally do not want to do.

I know I do the same thing, but at least I enjoy it. If one fine day I decide to pack up and quit, the ocean depths are still open to me even if the surface isn't. But I know I'll never do it. Your famous "domesticating", carried out to perfection on my father and wife, has had a notable effect on our make-up. In many ways I have passed this on to my own young. Despite no change in our physical appearance, have we become half animal half human? I don't know, but just the idea of it really bothers me.

When I discuss these things with my wife, our children stare at us in amazement to hear things they don't understand and will probably never understand. This embarrasses me and we end up in silence because we don't know what to say next. Eventually I break the silence by leaping twice as high as I usually do in the performance and the whole family follows suit to compete with each other.

Our human teacher joins in our laughter like one of the many children who visit the aquarium.

We must always be loyal to each other. Your friendship may be open but it is demanding. If one day we really did summon up the courage to leave and go back to live in the ocean with the other dolphins, I fear that they would be worried and afraid and suspicious of us.

As though we were strangers.

THE WOLF

Dear Humans,

know that many of you are overjoyed about the fact that all over the world we wolves have been reduced in number and are on our way to extinction. However, I am also comforted to know that most of you are saddened by this. It's a sign that the myth about the wolf's legendary ferocity is at long last changing. But you probably won't fully understand the role we play in the increasingly more precarious ecological balance of our shared "mother earth" and animal life in the universe unless we wolves were to completely disappear.

Continuing to write the same old stories about wolves and our blood-thirsty cruelty is unjust. Your saying that a human is "as ferocious as a wolf" is pure superstition and serves no purpose. What meaning could slanderous nonsense such as this possibly have for a people who have already landed on the moon?

This is definitely not the "era of wolves". It is the human era. Perhaps even that of the more frequently appearing wolf in sheep's clothing. It's about time your psychological, genetic and behavioural knowledge explained to us wolves and yourselves something about our presumed and proverbial aggression which has been intentionally utilized, even in the most innocent of fables (now seen to have Freudian connotations and grimly exploited).

I only refer to the "wolf in sheep's clothing" so you understand what I'm trying to say.

I have always dreamed of being your friend. When the dog came into the picture and replaced me, it only built up my fine hopes even more. If you were to stop using me as the ferocious "baddie" and began putting an end to the countless number of fictional roles I am playing, we could have the pleasure of each other's company in the long journey back to friendship on earth. You who devour everything ever written in illustrated magazines and learned articles about planets, stars, geology and animals, about the earth and everything extraterrestrial, you should know that you have nothing to fear from the wolf.

We are never ferocious, we never attack humans and very rarely any other smaller or weaker animal unless we are hungry or desperate. We have managed to survive among people in so-called civilised countries. Only in exceptional cases, never as a matter of course, are we forced to attack and kill. But, at the very most, this is only a raid on a hen-house or a flock of sheep. The slightest flicker of a flame is enough to make us flee in terror. Did you know that we are ashamed of being so afraid of fire? And equally ashamed of the rare occasions when starvation forces us to attack you, perhaps at sunset when you are isolated in uninhabited woods or other places where you are vulnerable to our superior strength? And did you know that your flesh and blood are far from being our idea of a tasty banquet? We much prefer your chickens, rabbits and sheep.

Why can't we sit down and talk about this together in the language spoken by Adam and all the animals in the Garden of Eden? We could clear up all sorts of obscure problems, both old and new. And really become friends again. Just think, we could play together as we used to in that garden. For now, the only one privileged enough to play with you is my counterpart, the dog (heaven help me if he's listening!). For some reason, things suddenly went very wrong in that garden and everyone became carnivorous. As animals became the staple food of the human diet, human flesh was also eaten by animals. We all became each other's prey without exception. The human won the day when he defined indiscriminate hunting of us all as a "noble art".

However, the world, the earth, air, water, light, flowers, day and night, all these things have remained more or less the same. Despite the devastation, they have maintained a delicate and fertile balance. Only in the last century have you humans – and only you, there's no question – become like crazed savages, bent on poisoning everything into a barren wasteland.

We "ferocious wolves" are also present and caught up in this precarious balance. You can read about it in any encyclopaedia. They tell you about it at school. The more honest scientists and ecologists can prove it to you. You understand, agree and perhaps even get a good mark in the subject of natural science. Then you grow up and get off course, indifferently tearing that balance to shreds. You make a full-time job or even hobby of ripping up any beauty, wisdom, truth or loyalty left alive and unharmed, untouched and magnificent.

Yes, I know, you make a public display of combatting for "biological balance" and dozens of other equally worthy causes that have now become a cornerstone for the evolved, up-to-date human status symbol. But if you claim to know all of this, why do you keep picking on our allegedly ferocious and violent nature? You should be ashamed of yourselves for using us in your cartoons and horror films (regurgitating big, bad wolves, werewolves, demons born from a black she-wolf...). There's a wolf in almost every fairy story. He's always bad and always loses in the end. The one who tried to get his sticky paws on dear Little Red Riding Hood is, of course, the most famous.

Brother human, why don't you think about your own massacres first? They are far more widespread and better documented than ours. They are tolerated, sometimes acclaimed, even blessed and immortalised in sincere and naive "ex voto" attached to the plaster clothes worn by statues of the Virgin Mary or St Francis. Don't stuff us and present us as ferocious animals in your horrible museums.

Even if it isn't exactly a bonus (there's no escaping the "big and bad" bit), being the star of a fairy tale like Little Red Riding Hood could be thought of as quite thrilling, except of course for what I have just said. Despite its well-established ambiguity, it continues to be a sound, indestructible story, tried and tested generation after generation. However, as it seems to have been written more for adults rather than children, it's not what you might call educational. As for the rest, you must stop referring to the "wolf" at every chance you get. The slaughter that has already taken place as a result of crimes committed by wolves in sheep's clothing show him to be a disgraceful dictator, a military fanatic, a tyrant or inquisitor. Today he would be called a terrorist or a killer "monster" of young courting couples.

You'll probably tell me that I'm repeating myself, but I'm doing it on purpose. Think of the bomb, the Bomb with a capital B, the thing that takes a second to wipe out everything on the planet and beyond, every trace of memory and history. That Bomb really is a Big, Bad Wolf. And you gave birth to it. If you're looking for excuses, I dare you to compare yourself with the most quoted and loved of all wolves in Christian history and literature, the Italian wolf of Gubbio, and not with your nordic myths or legends like Romulus and Remus, suckled by one of my ancestors. Did you ever consider that we know that story better than you do? We gladly hand it down, generation after generation. It is a marvellous story, a prophecy and hope that has burned for eight hundred years in the furnace of our reconciliation. This story not only honours that pauper whose wealth of grace and friendship, peace and joy enriched both the human and animal world, but also pays homage to our forefather the wolf, who only became ferocious because he was half-starved. He found Francis to be not only an extraordinary friend but a severe judge, a person who persuaded him to "mend his ways" so that he could see how those humans whom he terrorised could look after him like a brother, a well-loved pet dog. Until his peaceful death.

With a bit of persuasion from Francis, those humans and the wolf came to terms. Instead of generalising with a "let's love one another" they drew up a specific peace treaty of non-aggression and permanent collaboration. That pact was drawn up before the entire population of the city. The "Fioretti" stories written about the life of St Francis tell us that "not even a dog barked" at the wolf who was once an enemy.

I can assure you that the very same wolf, in whatever year it may have been in Gubbio, never left the forest, never approached Francis with bent head and never made the raids of which he was accused. And you know why? Because that wolf never existed. It was not a wolf. It was a human, a wolf in sheep's clothing. Perhaps one of the last aristocrats to be thrown out of the city when Gubbio became a democratic municipality, living like a desperate and starving outlaw, just as a real wolf does. Those exiled people were so ferocious that they carried out raid after raid both in the city and the countryside. Perhaps because the others had already died of starvation and privation, in all probability that desperate survivor fed on the flesh of the many wolves abounding in those days in the Gubbio woods. In order to steal and kill, it would seem that this beast-like human wore a mask made from the head of a wolf which he had killed and gradually eaten. This is probably how the legend came about.

Or is it just an unfounded supposition? Maybe. I couldn't swear to it. But if you do accept it, this supposition honours the human more than a real wolf. If I'm not mistaken it was the human who accepted the invitation given by Francis, the impoverished mediator, to shake hands (or paws) and say yes to that peace treaty. Surely you understand that these are human qualities. Just think of the amazement and the mixture of terror and happiness when the saint called the human who then came out of the woods, removed his mask and revealed his true identity. Knowing us and loving us as he did, Francis also knew that he was not insulting us when he related the episode to the monks who were not present: "My brothers, if you tell this story to your descendants, don't humiliate this human by mentioning his name now that he has mended his ways and become a friend to one and all. Say that he was a wolf. There always have been and always will be ferocious wolves in the woods around Gubbio. None of them will take offence. In time everyone will understand the truth. Everyone will feel joy, no-one will be ashamed."

However, dear friend, eight hundred years have gone by since that story took place. Out of pity for you, Francis was the only one who could rightfully entrust us with the role of standing in for the human. You no longer have this right. You are humans; we are wolves. We don't want any confusion or lies. A ferocious human is a ferocious human; a ferocious wolf is a ferocious wolf. Hunger is the same for everyone. If there is no justice and fraternity dies, everyone, both humans and wolves, becomes a beast. We must all assume responsibility for our own violence, our own sins and our own penitence.

I hope you understand and try to get to know us better. Although not in keeping with widespread opinion, with any luck you will discover that, in our world, the male wolf is a most loving, tender and loyal father. If necessary, his ability to provide everything and attend to everything his cubs could ever need is just as great, if not greater, than the female's.

Here's my hand – or my paw – extended to you, the wolf in sheep's clothing, my brother. Please shake it. Put a stop to your old fears that make you "cry wolf".

THE HORSE

Dear Humans,

It didn't take long for both humans and horses to forget that we probably appeared on earth at the same time. Nonetheless, a sense of equality has always remained in our relationship. For me, this is a source of great pleasure. No other relationship between human and animal is as complete, stable, dignified and respectful as ours. Believe me, no-one respects the human as much as the horse does. I also believe that no-one loves and takes pride in the horse as much as the human does. Neither of us feel superior to the other. Quite a miracle, don't you think?

However, as you know, the biogenetic origins of our species are only important up to a point. The "elective affinities" which have always joined us are far more important. The rumour that we are cast from the same mould probably came about as a result of our deep affinity.

Despite the charm of poetry and visual images depicted by day and by night in ancient mythology, especially Greek and Mediterranean mythology, you must put the centaur into perspective. The centaur did not start out as being one creature, and, if it was, it became that way through the gradual and perfect fusion of two creatures, the human and the horse.

That charm and those visual images have been used time and time again throughout ancient art. In many ways, this charm is still a source of marvel. Every time a human who knows and loves me uses his intelligence, friendship and common sense to ride me, we join once more to become one creature. At times like these, there's no need for words. More often than not, total silence works best. We are on the same wavelength and it's not unusual for us to achieve balance and harmony through our thoughts.

I, the horse, am an indispensable part of your history, civilisation and culture. Before the machine age, I was your strongest, best, most efficient and intelligent means of transport. The war-horse is well documented in old reports of human slaughter. Above all, however, he is your fraternal companion in peace, enjoyment and competitive sport. Not to mention your shows. Until very recently, your children have been more excited by horse shows than any other. They still go even now, despite their addiction to the violence of computers, robots and toy spaceships.

We share a lot of memories, don't we, dear human? What would the first motion pictures have been like without horses? The "western" would never have been created, nor would silent movies have attracted so many to watch the historic epics with their chariot races in Nero and Cleopatra's arenas. All over the world, books as important as your *Black Beauty* have been written for me and about me. I have performed in front of your movie cameras. Attached to those contraptions in pioneer wagon-trains, I have raced, taken terrible falls without hurting myself or others, scrambled to my feet and always carried the hero to safety. I have galloped across thousands of moors, prairies, rivers and deserts. I have acted roles in love stories, power struggles, droughts and famines, even death. I have relived every part of your foolish or marvellous human adventures, from the Homeric Wars to the migration of the Pilgrim Fathers.

I have guarded my dying rider and galloped back to the ranch to get help to save him.

Then, unfortunately, came the boom in race tracks and betting shops run by the Mafia. Even though I am forced into it, I am very glad that no-one dares accuse me of being connected with this corruption. I am more than content to be remembered in sculpture, bas-reliefs, paintings and friezes of heroic glory. All these documents prove that not one of the greatest civilisations has ever been able to do without me.

Without me, only half of these civilisations would have existed. Perhaps some of them wouldn't have existed at all, or been untraceable. Half of the beauty, the power, the commanders' glory and immortality belongs to me. Without me, how could they have been sculpted in marble, cast in bronze and glorified in poetry? No blame can be attributed to me for the actions of certain "leaders", for want of a better word, and mass murderers. I have been declared innocent and a positive force throughout history and civilisation. Unfortunately, equestrian monuments have cropped up all over the place during recent centuries. They are unbearably ugly. I also lose by this and it does no please me at all.

Do you think I am boasting too much?

I am not boasting, just making some observations. And I still feel the strong desire to carry on living with you, but in friendship, for peaceful endeavours, for adventures which keep your body in trim and lift your spirit, tone up my muscles and make me happy.

Your reference books tell you that wild horses are almost extinct, or have been extinct for some time. But don't you believe it. Fortunately some wild and free blood still runs through my veins and somewhere, in the Italian marshes or the American prairies, the horse still lives with a degree of freedom. These are the ideal conditions for renewing the friendship I mentioned before. In the "rodeos" so often shown in past films, you always depicted us as proud and untamable, bucking off inexpert riders of all ages or clowns. This is not the right image at all. I don't want a tamer. I want an equal, a human who can equal me in ability, discipline and freedom. The less ignorant and presumptuous among you are aware that this is the best way to renew our bond and live together. Given, of course, that we both survive all the pollution. You know that a gentle stroke and a lump of sugar is all that's usually needed to face our strength and agility. What you don't know is the extent to which your friendship and respect keeps us going, nor how much we deserve your admiration.

As I said before, more often than not we fully understand one another at a glance, through a soft murmur, a gentle stroke or even in total silence.

Our affinity also depends on the length of time we live together. A horse lives with his children and often also his grandchildren while continuing to live with his father. Although the horse does not live as long as the human, his life is still quite long and during so many years he learns to know and love the human and reciprocate in all things. In a life span of twenty to forty years, I have, so to speak, time to become naturalised, at one with you and your family. This is a privilege no other animal can dream of having. We have all the time in the world to learn from each other. I'm not in any hurry. I can afford the time and pleasure to watch how we both sort out our lives.

Let's take a look at this life, this historical and legendary adventure. It's a unique story and one of the most marvellous still being told. To many you are a hero, and so am I although all we have is physical strength and an instinct for things invisible and words unspoken. But let's view our present life as an experiment in making up for past mistakes and arrogance, violence and whims. Without turning our backs on the past, let's try to live it a new way, differently, by salvaging something from our ancient history.

I wish the day would return when the word "cavalier" meant a horseman and the best form of praise for me, the horse. A horseman is a grand "gentleman" who treats everyone properly. Be it a factual story or a fictional epic poem, I am pleased when the phrase "that horse had a human expression" is written. Despite being full of airs and graces, the word "equestrian" is a term which could lead us back to our common origin. I just wish that "equestrian" would cease to be applied to monuments of humans riding horses. These are neither humans nor horses. It's only a monument, a hundred, a thousand monuments...

But I'm just day-dreaming, and I'd better get back to the point.

Like the many other species of animal who are dying out, if not extinct already, I too have been defeated. This is the truth of the matter, dear humans. At this rate, I don't just fear the defeat of the "wild" horse, but of all species of horse.

You may admire us, nevertheless, with your pride bordering on sheer stupidity, it is you and your automobile industry who have willed our disappearance. We've almost reached the point where your children will only be able to see me in films. During the summer, with families dashing off for a quick day-trip into the increasingly more spoiled countryside, your children may come across a wealthy man whose good taste permits him not only the luxury of a sports car but also a horse. And your children may have difficulty recognising that strange animal as the horse they have seen in films. The image will be soon forgotten however, as soon as they've crammed in their quota of relaxation and have turned back home to face the traffic jams of Sunday drivers moving towards the cities at a "snail's pace". Thank goodness you don't refer to it as "at a horse's pace"!

The streets and city squares where I lived as king for thousands of years are no longer mine. We sense the fast approaching end of our species as we, the last of our kind, watch the daily increase in motorised tin cans belching their fumes over the sweet face of the earth. Our end will not occur as a result of natural extinction or evolution of the species in Mother Nature's difficult yet sweetly mysterious grand scheme, but because of the brutality of intensive and pitiless technological progress.

Quite simply, there is no more room for us.

And yet, dear humans, I won't give up.

Although I'm aware that the odds are against me, I still hold out hope.

We must try to turn our dreams into reality, make them come true.

If you'll agree, I'd like to to see if I can turn into a hippogriff (that's right, I'm that desperate I'd even become a winged horse) and try to gallop to the moon (alas, you've already been there, and without the horse). A hippogriff, the horse of visionaries and poets like your Italian, Ariosto, must be better than a car or a spaceship. I have, in my dreams that is, often been a winged horse. But, even with wings, I felt less of a horse than before because you weren't there.

We mustn't give up. One way or the other, we have to find a short-cut, a gap in the snaking queue of tin and noxious fumes blocking the roads and poisoning the countryside, the beaches, woods and rivers, breaking the silence and disturbing the peace.

Let's try! Jump on my back and let's be free at last,

before we die together, to gallop along the shore at sunset, poised between the earth and the sea, never to stop again. We will be as happy as we were on the earth's first morning.

Before the sun ceases to rise forever.